*Guerrilla
Marketing
Attack*

GUERRILLA MARKETING ATTACK

New Strategies, Tactics, and Weapons for Winning Big Profits for Your Small Business

Jay Conrad Levinson

Houghton Mifflin Company
Boston

For information about permission to reproduce selections from
this book, write to Permissions, Houghton Mifflin Company,
2 Park Street, Boston, Massachusetts 02108.

Library of Congress Cataloging-in-Publication Data

Levinson, Jay Conrad.
  Guerrilla marketing attack.
  Includes index.
  1. Marketing. 2. Small business—Management.
I. Title.
HF5415.L476   1989   658.8   88-26670
ISBN 0-395-47693-3
ISBN 0-395-50220-9 (pbk.)

Printed in the United States of America

S 10 9 8 7 6 5 4

For my own Magnificent Seven:
Patsy, Amy, Sage, Sadelle,
Myrna, Gina, and Nikki

Preface

This book is a sequel to *Guerrilla Marketing*, which describes a new way of marketing your goods or services — relying on your time, energy, and imagination rather than your bottomless marketing budget (which you most likely don't have anyway). If you haven't read *Guerrilla Marketing*, you'll probably want to after you finish this book. The information in that book will provide you with a solid foundation for marketing any kind of business, especially a new, small, or medium-sized enterprise. *Guerrilla Marketing Attack* is more than an update on that crucial information.

Guerrilla Marketing Attack is a method for using the newest insights of guerrilla marketing to plan, launch, and maintain an all-out marketing offensive designed to transform prospects into customers and marketing investments into profits.

The Department of Commerce alerts us to the fact that seven out of ten new businesses won't be operating after five years. This is alarming news. Things are only going to get tougher out there as more and more disenchanted former jobholders opt for the enchantment of their own business. To survive as a small or medium-sized business, you're going to have to be a guerrilla.

More companies than ever will be marketing to your prospects. Marketing will take on increasing ferocity. You must do something about it. You can — and should — wage an all-out marketing attack. This book will show you how to achieve a victory money can't buy — using the new strategies, tactics, and weapons of guerrilla marketing to mount your own guerrilla marketing attack. It's not going to be a picnic, but it will result in success.

Bill Shear, President of Guerrilla Marketing International, is the man most responsible for my own mounting of the guerrilla marketing attack. Bill has masterminded my spreading of the word about guerrilla marketing throughout North America. He has been the guiding force behind our guerrilla marketing seminars and lectures, our guerrilla marketing consulting, *The Guerrilla Marketing Newsletter*, the guerrilla marketing audiotapes and videotapes, and many tactics of guerrilla marketing that are both used by our organization and taught to those who attend our presentations.

Acknowledgments are also owed, along with gratitude, to those guerrillas who have illuminated my way through the marketing battlefield: Allan Caplan, Michael P. Lavin, Robin A. Bacci, Bruce Kaplan, Jane Croston, Ellie Dasher, Kevin Shafer, Jack J. Freeman, Dale and Merl O'Brien, Mike Larsen, Robert Holden Marriott, and Vicki Gross.

To all these people, and the many unnamed heroes and heroines of marketing wars in which I have participated, I say thank you, and I urge you onward to victory.

Contents

Guerrilla
Marketing
Attack

1
One Hundred Weapons
for the Attack

YOU ARE SURROUNDED. All around you are enemies vying for the same bounty. They're out to get your customers and your prospects, the good and honest people who ought to be buying what you're selling. These enemies are disguised as owners of small and medium-sized businesses.

The enemy is disguised

Several of the enemies are grossly larger than you. Some have the power and personality of Godzilla. Many of them are far better funded than you. Some have been successfully operating their businesses since prehistoric times.

These enemies thrive on competition. They're out to get you and get you good. They're out for the disposable income currently held by your hot prospects and past customers. They're out for the attention of every red-blooded consumer who reads the newspaper, listens to the radio, watches TV, grabs a handful of junk mail out of the mailbox, or gets a deskful at the office.

Your enemies mean business: *your* business, *your* profits. Some of them can run more ads in more papers and more commercials on more stations than you'll ever run. They can mail more materials to more people than you'll ever mail. They can outspend you in every arena of marketing that money can buy. But they can't outspend you in marketing arenas that money can't buy. And they can't always outthink you. If you put up the time, the energy, and the imagination, you can gain the same marketing leverage that many of your enemies get by putting up megabucks.

What the enemy can't do

If you decide to live by new strategies and practice new

Outmarket the competition

tactics, you can get a substantial piece of the pie. If you begin to use a low-cost but power-packed arsenal of potential marketing weapons available to you, you can actually outmarket your competition. If you don't, at least one smart competitor will outmarket you. Marketing is emerging from its adolescence, and if you don't use it to attack in the battle for prosperous business survival, you're going to be the innocent victim of someone else's attack.

Not only is marketing gaining sophistication, but you are being beset by more enemies than you probably imagine. If you run a business that specializes in selling a service, you may think that your enemies are those who also sell a service just like yours. But that's incomplete thinking. Your enemy sells a service like yours, a service unlike yours, a product, or all of the above.

Enemies are everywhere

Your enemy might be a bank, an airline, a department store, a stereo system, a computer manufacturer, a telephone company, a jeweler, or one of the many businesses that sell only to other businesses.

Becoming a guerrilla

It's a marketplace out there. In order to survive, let alone thrive and prosper, you've got to be a guerrilla. There are ten requirements for becoming a guerrilla:

1. You've got to open your mind to the full extent of marketing. It's fuller than you thought.
2. You've got to adopt the personality of other successful guerrilla marketers. Without it, life will be tough.
3. You've got to think about marketing differently. A lot of the old truths have turned into myths.
4. You've got to plan your guerrilla marketing attack with an easy-to-understand, easy-to-follow battle strategy.
5. You've got to define what you want your attack to accomplish with precision and realism. If you're not defining, you're not attacking.

The first five requirements are just the fundamentals for winning the battle for healthy, honest, and growing profits. They will serve you well on your way to the battlefield.

6. You've got to attack — do exactly what your plan said you'd do. You must take action.
7. You've got to understand which media can best serve your needs.
8. You've got to orient everything about your business to your customer — that person above all who can help you prosper. Your customer must sense your dedication.
9. You've got to recognize the fast-changing nature of marketing today. It's wilder than ever, but you can master it if you keep up with it. Guerrillas do.
10. You've got to maintain your guerrilla marketing attack. You can fulfill the other nine requirements superbly, but if you don't fulfill this one, you're a goner.

The news from the battlefront is good

Don't think that launching a guerrilla marketing attack is all hard work and demanding requirements. It's a whale of a lot of fun winning battles. Your competition isn't half as savvy as you'll be by the time you're a full-fledged guerrilla. About half of the marketing weapons available are absolutely free of charge. And the task of marketing will get easier and easier the more you do it.

Marketing gets easier

Chances are, your direct competition has quite a bit of marketing experience. That's probably going to work in your favor because you can rest assured that marketing changes faster than your competitors do. Many of your enemies will be bogged down in myth and tradition.

Understand that most of your direct competitors are living in another world when it comes to marketing. Except for the sophisticated and well-capitalized few, the majority of businesses competing for the same dollars as you labor under archaic marketing principles that have been out of date for over a decade.

You are about to be presented with a treasury of 100 guerrilla marketing weapons. It might be difficult to suppress a grin when you realize that *the majority of your direct competitors* are aware of no more than twenty or thirty — more likely ten or fifteen — of the 100 marketing weapons you can use. And chances are, they're using only about three or four. Even the hotshots aren't using many more than ten.

Here are scores of ways you can outmarket the hottest of those hotshots, while achieving business longevity and a lengthy customer list. Your bank account will brim with profits in direct proportion to how your marketing arsenal brims with these weapons.

The 100 guerrilla marketing weapons

1. *Name:* There are zillions of good company names and zillions of bad ones. Just be sure yours is one of the good ones. Be sure people can pronounce it, that it does not confuse them, and that it is uniquely yours.

Positioning
2. *Product or service niche:* This is also known as *positioning* by those who love buzz words. It refers to the segment of the market you want to call your very own.

3. *Color:* Memorability is increased and attitudes established when you associate a color with your business. You'll use that color in many of your marketing weapons.

4. *Identity:* This is not a false image, but an honest identity that conveys your company's personality throughout your marketing. Be sure it realistically reflects who you are.

5. *Logo:* Some people call this a *trademark* or *symbol*. Whatever you call it, it's a graphic representation, more visual than verbal, of your company. It's smart to have one.

6. *Theme:* This is a set of words that summarizes your company or its prime benefits. Pick a theme you can live with for a long time. The longer you use the theme, the better.

7. *Package:* Your package is the box your product comes in, the office your services come from, your van, your sales-

people, you. The way you package your offering will attract or repel your customers and prospects.

8. *Size:* The size of your business influences some people to buy or not to buy. Big is not necessarily good. Neither is small. But both can be good. Can you offer the benefits of both?

9. *Decor:* Prospects and customers will form opinions about your offering based on the decor of your store, office, or factory. It should reflect your honest identity.

10. *Attire:* You and your employees represent your business. People will develop attitudes about your business based on what you and your employees are wearing at work.

11. *Pricing:* Pick a pricing niche — high, medium, or low — depending on your competitors. Among consumers, 14 percent say price is the main determining factor of whether or not they will buy. **The pricy 14 percent**

12. *Business card:* Guerrillas make theirs more than a name, address, and phone number. Print the benefits of doing business with your company and turn your card into a minibrochure.

13. *Stationery:* The look and feel of your stationery make it a powerful marketing tool. Millions might see it if you do direct mailings.

14. *Order form/invoice:* This is not just a business form, but an opportunity to gain more business, increase referrals, and solidify the relationship between you and your customer.

15. *Inside signs:* These spur impulse purchases, act as silent salespeople, merchandise your offerings, and cross-merchandise — all at the same time, and inexpensively, too.

16. *Outside signs:* They may be near your place of business, far away, or moving about on a bus or taxi. They direct people to your product, service, or store.

17. *Hours of operation:* Yes, your hours are part of your marketing. If you're open evenings (or early mornings) and your competitors aren't, you may gain business from them.

18. *Days of operation:* I know of a business that decided to

stay open Sundays because nobody else did. Soon, Sunday became their second most profitable day of the week.

19. *Phone demeanor:* How you answer the phone turns people on — or off. You and the people who answer your phones should know exactly how to turn on each caller by words and attitude.

20. *Neatness:* Messiness causes many a lost sale because people believe the sloppiness will carry over into other parts of your business. Happily, the same is true of neatness.

21. *Location:* Without question, this is one of the most important parts of success for a lot of businesses. But it is only one part of marketing. There are 99 other parts, too.

22. *Window displays:* They should have high visibility, be unique, and include items that invite people inside. Change them regularly, and recognize their importance.

23. *Business plan:* This is like a map that is consulted regularly to be sure you are heading in the right direction. A good business plan has a marketing plan built into it.

1 percent of marketing

24. *Advertising:* Here's one of the most crucial parts of marketing, but it's only one one hundredth of the process. Some people think marketing means advertising.

25. *Distribution:* This refers to the ways your offering can be purchased: in your store, in the stores of others, by mail, by phone. Gain as wide distribution as possible.

26. *Service:* Consumers consider service or lack of it to be one of the four most important influences in selecting a business from which to buy. Service wins and loses customers.

The key to customer loyalty

27. *Follow-up:* Nonguerrillas think marketing ends when they've made the sale. Guerrillas know that's when marketing begins. Follow-up is the key to a loyal customer base.

28. *Customer recourse:* Know what you'll do if the customer is not satisfied. Have a clear policy so that disgruntled customers can be converted into satisfied repeat customers.

29. *Community involvement:* The closer you're involved

with your community — local and industrial — the higher your profits will be. This takes more time and legwork than money.

30. *Tie-ins with others:* Capitalize on countless opportunities to display the signs or circulars of other businesses if they'll do the same for you. Many will, cutting your marketing costs.

31. *Public relations:* This is publicity in the media based on something newsworthy about your business. PR is an important weapon in any marketing arsenal but not the only one.

32. *Publicity contacts:* The media are inundated with re- **The key to free** quests for free publicity, so if you have contacts in the **publicity** media, the chances increase of your publicity appearing.

33. *Reprints of ads and publicity:* Most publicity stories appear but once; most ads are costly. Gain maximum mileage from both by making inexpensive reprints for mailings and signs.

34. *Special events:* Staging unusual activities around your business is a good way to attract free publicity. Have a contest and award a trophy; throw a party for prospects.

35. *Testimonials:* These are free, easy to obtain, and very impressive to new prospects. Use them as signs, or in your brochure, ads, or direct mailings. They work.

36. *Smiles:* You won't find this mentioned in marketing text- **Smile when you** books, but a smile is a part of marketing and makes your **say that!** customers feel special. Employees should smile in person and on the phone.

37. *Greetings:* The way you say hello and goodbye offers still another chance for you to single out each customer. Your warm greeting will be translated to word-of-mouth marketing. So greet warmly, using a smile, eye contact, and, when possible, the person's name.

38. *Contact time with customer:* Every moment you're with the customer is a marketing opportunity. Use it to intensify your relationship, market other items, be of better service.

39. *Sales training:* The more training you do, the more prof-

its you'll earn. Be sure all salespeople know the key parts of the way you do business and reflect your identity.

40. *Sales presentation:* The best salespeople use certain words and phrases. This indicates that the best sales presentations should be memorized, then delivered with zest.

41. *Sales representatives:* These people will deliver your presentation; whether they're employees or independent reps, be sure they see your business the same way you do.

42. *Audiovisual aids:* Give your sales reps and your place of business the benefits of visual aids. Points made to the eye and ear are 68 percent more effective than points to the ear only.

68 percent more effective

43. *Audiotapes and videotapes:* Use these to establish your expertise. They are electronic versions of brochures; many guerrillas can use them inexpensively and effectively.

44. *Refreshments offered:* Little things such as offering coffee and doughnuts in the morning can have a dramatic impact on sales. Also consider apples, candy, juices, wine, cheese.

45. *Credit cards:* The easier you make it for someone to buy, the more they'll buy. It's worth the percentage and the paperwork to accept many credit cards and gain many sales.

No money is no problem

46. *Availability of financing:* Many customers will want what you're selling but won't have the money now. Letting them know they can pay later can win sales otherwise lost.

47. *Club and association memberships:* Join these to become part of the community and to make friends. But be sure you also join them to work hard and earn your new friendships.

48. *Team sponsorships:* One more way to involve your business in the community and to meet potential customers. If you want them to buy what you're selling, show up for the games.

49. *Word of mouth:* You can control this by providing superb

service, informative brochures, and steady mailings. If you establish a bond with your customers, you can even ask for referrals.

50. *Circulars:* These are inexpensive, flexible, and easy to distribute on street corners, under windshield wipers, as bag stuffers, as signs, on counters, at homes, and more.

51. *Brochures:* These are more detailed and less time-bound than circulars. Their purpose is to provide overall information; they work well as a free offering in a small and inexpensive advertisement.

52. *Samples:* If you have enough quality, sampling is one of the most effective marketing tools ever devised. The most knowledgeable marketing companies in America use it.

The power of a free sample

53. *Consultations:* These are like free samples and work very well for service businesses, especially information services. Free consultations also help sell many products.

54. *Demonstrations:* Once again, demonstrations let prospects see what it would be like to own what you sell. Demonstrations are what make TV such a powerful marketing tool.

55. *Seminars and lectures:* These forums not only allow you to establish yourself as an authority, but serve as the springboard to the sale of both products and services.

56. *Column in a publication:* Many local and business publications will publish a column on your field of expertise. Don't ask for money, only for mention of your business name and phone.

Fame, not fortune

57. *Books and articles:* Credibility is an obvious aid in marketing, and these published pieces increase your credibility. Even self-publishing will establish credibility.

58. *Contests and sweepstakes:* These call attention to your business and obtain precious names for your customer mailing list. If possible, have entrants come to your place of business to enter.

59. *Phone-hold marketing:* When all your telephone lines are busy, this answering machine puts your callers on hold, then plays music while imparting useful information about your company.

60. *Music theme:* If you own the rights to a piece of music, you can use it as your theme for your answering machine, radio spots, TV commercials, and audio or video cassettes. Listeners will soon identify the theme with your company.

61. *Booths for malls/streets:* These inexpensive and portable structures can give you an additional location in a hurry. They also may be used in many places that can be highly profitable.

62. *Roadside stands:* Perhaps your product can be sold from your own roadside stand or from the stand of another vendor. These marketing tools should be considered if you sell products.

63. *Farmers and flea markets:* You can increase your distribution, not to mention your sales, if you sell or have reps sell your products at these regular large gatherings.

64. *Access to advertising materials:* If you sell products or services made by others, perhaps they have expensive advertising materials they'll furnish you for free.

65. *Access to co-op funds:* Many manufacturers make funds available if you give them a plug in your ads. In 1986, over $2.5 billion in co-op funds went unclaimed. Claim yours!

66. *Research studies:* The more you know, the better you can market. Your industry publication or reference librarian might be able to supply important data on your industry, market, and prospects. Or do your own research.

67. *Classified ads:* These are widely used by small businesses throughout America. Consider using them in local or national newspapers, also in national magazines. They are inexpensive to test.

68. *Newspaper display ads:* These ads are the prime marketing medium of the small American business. Build an inventory of newspaper ads and offers that have worked for you.

69. *Magazine ads:* These lend to your credibility. Many important national magazines now publish affordable regional editions. Consider both trade and consumer magazines.

70. *Yellow pages ads*: If your competitors aren't there, you don't have to be there. Otherwise you do. Be certain to use these ads to give as much information as possible.

71. *Direct-marketing coupons*: Companies compile coupon cards into decks called postcard decks, then mail a group of related coupons for products and services to target groups. The mailing cost is shared by all who supply the coupons.

72. *Direct-mail postcards*: Postcards are valuable weapons in the quest to break through marketing clutter, cut costs, and keep up steady communications with customers.

    Breaking through the clutter

73. *Direct-mail letters*: You'll be able, after enough testing, to know which letter will produce known results from a specific audience. Each letter should be self-contained.

74. *Catalog*: When you have ten thousand or more names on your mailing list, consider a full-color catalog. If you don't have enough names, you might try a two-color catalog.

75. *Newsletter*: Newsletters serve as a regular mailing tool, establish your expertise, serve as an efficient marketing medium, even can be a profit-center of their own.

76. *Inserts*: Inserts are 4-to-8-page brochures that are inserted into the daily paper or mailed. They are an effective form of direct marketing because they enable you to reach every family in a zip code.

    Targeting zip codes

77. *Trade-show display*: Some businesses gain all the sales they want by means of a professional trade-show display that dramatically demonstrates the benefits of their offering.

78. *Merchandise displays*: Large or small in-store displays of your offering and its benefits often make the difference between gaining or losing distribution and sales.

79. *Billboards*: These are seen by large numbers of motorists and do a good job of telling or reminding people of you. Keys to their success: great locations, short copy.

80. *Balloons, blimps, and searchlights*: You've got to find ways to make yourself distinct from your competition. These are among the ways. They also help boost foot traffic for a special sale or special event you are holding.

Help folks remember your name

81. *Advertising specialties:* There are myriad gifts that can be imprinted with your name, reminding customers of you and nudging them to buy from you. Calendars and scratch pads are some of the most common ad specialties.

82. *Posters:* These may be blow-ups of your ads, blow-ups of elements of your ads or brochures, or anything you like. They lend color, pizzazz, and visibility to your identity. Display them in stores, airports, offices, and elsewhere.

83. *Bus and wind shelters:* You can put your poster or posters in these practical shelters that protect people from the weather, and market your offering with art, copy, and verve.

84. *Telemarketing scripts:* These scripts are used by people making sales phone calls for you. They contain key sales presentation ideas, questions, and even ways to close.

85. *Take-one boxes:* Place one of these in any location frequented by your prospects, fill it with your brochures, and watch how many people read what you have to say. It's high impact at low cost.

86. *Radio commercials:* These are usually thirty or sixty seconds long, may or may not have background music, and are part of a campaign carrying the same basic marketing theme. As a weapon, these should be in prerecorded cassette form.

The undisputed heavyweight champion

87. *Television commercials:* The most powerful marketing device in history lets you sell your product or service with words, pictures, and music. A potent weapon, the commercial shows the product or service in use and demonstrates its benefits. Again, this weapon should be in videotape form.

88. *Gift certificates:* Consider if your offering can be given as a gift. If so, these sales builders should be given serious thought. They work in many consumer-oriented enterprises.

89. *Gift baskets:* People enjoy purchasing entire packages. If you can imaginatively put one together, you can improve your overall sales without increasing your marketing cost.

90. *Human bonds:* These are relationships that transcend that

of mere buyer and seller. They are ties of humanity that create customer loyalty, indeed, even business reputations.

91. *Competitiveness:* This refers to your willingness to devote time and energy utilizing as many of these marketing weapons as are possible, sensible, reasonable, and affordable. **Competitiveness counts**

92. *Convenience:* Make it easy to buy what you are selling. Be easy to find and call and order from and pay. Do all you can to make it simple for customers to do business with you.

93. *Speed:* Time is more valuable than ever. People resent slow treatment more than ever. So they appreciate speed in handling their order or exchange or special request.

94. *Reputation:* This, more than any other single component, will turn prospects into customers. If you have a bad reputation, no great price will help you overcome it.

95. *Brand-name awareness:* With the phenomenal number of new business start-ups, those businesses that have built an awareness of their brand name have the best chance of success—now and in future years.

96. *Credibility:* If you have credibility, people will believe in your quality, your values, and everything you say about yourself in your marketing. Do anything to earn it.

97. *Enthusiasm:* This contagious marketing weapon gets passed on by you to your employees, from them to customers, and from customers to more customers. It's a healthy contagion.

98. *Customer mailing list:* Keep it from the day you go into business. Or start tomorrow. The longer your list, the higher your profits. This list is worth more than its weight in gold because of repeat patronage and referral business.

99. *Satisfied customers:* Your most powerful marketing ally is a customer who liked your quality and value. These people will singlehandedly create your word-of-mouth campaign. **Your most powerful ally**

100. *Marketing savvy:* You show some of it already by un-

derstanding marketing. You can show more by launching a guerrilla marketing attack. Marketing savvy means action.

Of these 100, fifty cost money; fifty cost no money whatsoever. Of these 100, the addition of one or even two new weapons to your own arsenal can make a dramatic difference to your bottom line. Think of the effect of eleven or twelve new weapons!

The ten most important weapons

If I had to make a list of the ten most important of the 100, I'd list these ten in a one-way tie for first place:

- Competitiveness
- Human bonds
- Credibility
- Enthusiasm
- Customer mailing list
- Advertising
- Reputation
- Service
- Brand-name awareness
- Satisfied customers

Let me be candid with you so that there are no surprises. Someday you might hear me give a lecture or conduct a seminar on guerrilla marketing. In that presentation there's a very good chance that I'll rattle off a hundred components of guerrilla marketing. But they won't be the same hundred I just told you about. That doesn't mean I'm inconsistent. It means that marketing is changing and I want guerrillas to be as up-to-date as possible. If that means updating my information, update it I will. In fact, I publish a bimonthly newsletter — *The Guerrilla Marketing Newsletter* — one of the prime purposes of which

State-of-the-moment marketing

is to keep my readers informed about state-of-the-moment marketing. I do that so as to maintain both their combat readiness and the quality of my information.

Please understand clearly that you've got to offer quality in your product or service. The guerrilla marketing attack is no shuck and jive show. It's a solid attack based on firm principles — the first of which is that *you've got to be devoted to excellence in your offering if marketing is going to work for you at all*. If you offer that excellence, the guerrilla marketing attack will motivate more people faster than standard marketing to buy what you are selling. But if you don't, it will truly speed the demise of your enterprise. And if you're selling substandard goods, the word will get around.

If you'd like a lump for your throat, here's one: For every complaint you hear, there are 26 other complaints you won't hear. Those 26 other dissatisfied customers will tell an average of 22 people each about the problem they had with your business. And 13 percent of them will tell more than 22 people. The word of mouth on poor quality spreads far faster than you'd ever worry it might. One person experiences it and the bad word travels like wildfire, only with more destructive power to **Spreads like** your business. You're probably insured against fire, but you're **wildfire** probably not covered against shoddy quality. So don't embark on your guerrilla marketing attack unless your offerings are excellent in quality and superb in value. Once they are, those offerings *plus* the guerrilla marketing attack make a potent combination.

The guerrilla marketing attack is more than mere book learning, and at some point you will have to give each of the 100 weapons some conscientious thought. When you've devoted that time — and I'm about to ask you to do just that — to figuring ways to use as many marketing weapons as reasonably and affordably possible to satisfy the specific needs of your company, you're well on your way toward becoming a practicing guerrilla.

Let's get down to the business of planning the framework for your guerrilla marketing attack, now and in the future. Here's what to do:

Readying the weapons for your attack

Begin by studying each of the 100 guerrilla marketing weapons and assign it to one of four lists:

List A: "I'm using this weapon now and I'm using it right."
List B: "I'm using this weapon now, but the way I'm using it can use some improvement."

List C: "I'm not using this weapon now and I ought to; I'll begin using it immediately."

List D: "This weapon isn't appropriate for my needs right now."

Look at list A. Love it, and continue doing all the things on it. The longer your lists A, B, and C, the better.

Next, look at list B. Lust for the opportunities it represents, then capitalize on every single one of them.

When to kick yourself   Now look at list C. Kick yourself for not putting these weapons into action earlier. Do it *now* — not the kick, the putting into action.

Finally, look at list D. Smile at it and put it away for the moment.

If you've got the personality of a successful guerrilla marketer, you're already champing at the bit to unleash some of your newfound attack power. But wait. First, let's see if you truly do have the quintessential guerrilla personality.

2
The Personality of the Successful Guerrilla

THE TYPICAL OWNER of a small or medium-sized business knows he or she ought to invest in marketing. But being typical, the owner isn't too sure about marketing in the first place, let alone the costs involved.

If that's you, deep down, you've probably always considered marketing to be somewhat akin to flossing your teeth. It's something you don't really want to do, but if you don't there's going to be trouble. So you do it. But you don't put much gusto into it.

Marketing is like flossing

After three decades in marketing, I wish I could tell you that the most successful guerrillas feel a sense of gusto about the entire marketing process. But I can't tell you that while looking you straight in the eye. The real truth is that I've known some very successful guerrillas who shied away from marketing as if it were the Black Death, delegating it with lightning speed to someone whose job description includes a gusto for marketing.

But whether or not all the successful guerrillas I've known had that burning desire to be involved in the marketing process, they all did — and still do — share five personality characteristics: patience, aggressiveness, imagination, sensitivity, and ego strength.

Since the first of these personality traits is the most important, it's a goal for me that you understand not only the word, but its fullest meaning in the context of marketing. To do this requires that I call your attention to a study where a group of researchers were asked, "How many times must a prospect see a marketing message to take them from a state of total apathy to purchase readiness?" Following a year-long study, the researchers concluded that a marketing message must penetrate the mind of a prospect a total of nine times before that prospect becomes a customer.

The power of nine

That's the good news.

The bad news is that for every three times you expose your prospect to your marketing message — via ad, sign, mailing, whatever — it gets missed or ignored two of those times. After all, people have more important things to do than to pay attention to your marketing. So you've got to put out the good word about your company a total of 27 times in order to make those nine impressions.

If you interpret that as meaning you might energetically market your product 26 times with a possibility of zero results, your interpretation is a good one, albeit depressing.

For example, if your primary marketing takes the form of newspaper advertising, and you run a good ad once a week in a newspaper — as I discuss and recommend in *Guerrilla Marketing* — here's how it generally works in real life for a new business that has done no marketing:

Cold reality

The first time: The first time your prospect sees your ad, he doesn't really think about it much. (This assumes your prospect is a he, but the reality is the same for a she.) Your message barely registers, but does get noticed. You've run it three times, but he's noticed it just once.

The second time: Now you've run your ad six times and your prospect is noticing it for the second time. He may read the headline, look at the visual, possibly even read the copy. But all that generally happens is that he thinks he may have seen that ad before. That's about it.

The third time: You've put out good money to run your ad nine times in a little over two months. But sales aren't soaring. Customers aren't pouring through the door or those tiny holes in your telephone receiver. What's happening? What's happening is that your prospect is seeing your ad for the third time and thinking he's pretty darned sure he's heard of you somewhere sometime. But another headline on the page grabs his attention away from your ad.

The fourth time: Now, you've run your advertising a full twelve weeks. Twelve ads you've paid for and your prospect has seen only four of them. Your prospect knows now that he's seen your ads before, and he figures that you must be offering something of value or else you wouldn't keep on advertising it. (Although people tend to believe that success breeds frequent advertising, it's interesting to note that just the reverse is true. I'll go into more detail about the effectiveness of advertising in Chapter 4.)

The fifth time: With your budget to the grindstone, you keep running those ads to those same people in those same media and your accountant still has no good news for you. The accountant wonders why you're putting all those bucks into fifteen ads without seeing tangible results. You wonder the same thing. But, at the same time, your prospect is beginning to feel a sense of familiarity with you. He's thinking of whether or not he needs what you offer. Maybe he should check you out someday. Yes, he'll do that.

The sixth time: Now, your prospect feels a sense of solidity about you. He's seen your ad six times over quite a long period — about a year, he figures, though it's only been about four or five months. He begins to think of how he can possibly buy what you're selling. Meanwhile, you've run your ads for eighteen weeks, and there's no real sign that they've taken hold yet.

At this point, most business owners figure they've been going about their marketing the wrong way. Maybe they shouldn't be running ads. Maybe the ads should have said different things. Maybe the ads should have run in different publications. Maybe the artwork was bad. So they decide to pull their ads — to stop advertising and look into something different.

Red light emergency! We can't let them do that! We can't let them stop their ads just when they're beginning to take effect. Can't they see that the ads are about to work? No, they can't see that at all. So they really do make a major and ill-advised change in their marketing, possibly eliminating all their advertising in the process.

This is the norm. This is the way it really happens, the down-and-dirty facts of life. The prospect is getting warmer and warmer about buying, but the advertiser is not seeing results. So the process comes to a grinding halt. I call this "sellus interruptus." The sale is never consummated. But consummation takes place when the advertising is allowed to run, as you will see.

"Sellus interruptus"

The seventh time: You have waited 21 weeks now, five whole months — and the advertising for which you have paid so dearly is not yet paying for itself, let alone paying for anything. Are you discouraged? You bet your bankbook you are! Who wouldn't be after 21 weeks of unanswered ads? Well, for one, the prospect isn't. Right now he's figuring out a time that he can own what you're offering. He's figuring how to pay for it, where to put it, even thinking about the benefits of ownership — the ones you kept reminding him of in your ads.

The eighth time: Well, you might sense a glimmering that the marketing is working by now. But it's not working that well. After writing checks for 24 weeks of advertising, you expected something more than a faint glimmer. Meanwhile, your prospect has it all figured out now — where, when, how, and why to buy what you've been offering. He feels good about you, feels he can count on you because you've been a constant presence in the media, because you've maintained a clear identity, and because he keeps reading about your company. Beware, for at this point, some business owners stop advertising. They just can't rationalize pouring any more money into something that's not working. So they just quit the media, quit advertising, break off all those budding relationships with new prospects. Those prospects, seeing the sudden end of advertising, have their confidence shaken, and decide to wait before making their purchase.

The ninth time: Cue the heavenly music. This is the time for people who love happy endings. The business owners who hung in there are delighted but not shocked when they see tangible proof of advertising's efficacy. Increasing numbers of prospects, people who were once totally apathetic to the business, knowing and thinking nothing about it, are now customers — real-life paying customers — and if you go about it right, repeat customers, the life force of a profitable business.

It took 27 ads to get through to these people. Nine times. They had to see your message nine times to gain, then sustain a desire to buy and own what you sell. Few business owners are willing to pay dues that high, wait periods that long, take nourishment from hope that elusive. But those who are willing to bear up under the pain of a lethargic sales curve gain the benefits of a valuable and rare personality trait: patience.

The first personality trait of the successful guerrilla is patience.

Of the five traits, if patience isn't the most important, then at least it's tied for first. What it's tied with may not be a word you like to use to describe yourself, but it's a personality trait you're required to develop if you're to launch a successful attack.

The second personality trait of the successful guerrilla is aggressiveness.

This characteristic is so important that it earns the first-place tie with patience. You can be a social wallflower. You can hide behind the potted palm at company parties. You can be unable to look even a kitten in the eye, but when it comes to marketing, you've got to be a *tiger*.

In the fight-to-the-death arena of marketing, you've got to think aggressively, spend aggressively, market aggressively. If your competitors don't hate you, at least they'd better fear you. They should fear that you'll outmarket them at every turn.

While they're making all the marketing mistakes, you're making none of them. While they're holding back on marketing, you're going all out.

In 1987, the average U.S. business invested 3 percent of its total sales in marketing. When a guerrilla hears that, the guerrilla says, "Three percent? It that all? What pikers! Think of what I can accomplish if I invest ten percent! Fifteen percent!"

That's aggressive thinking when it comes to marketing. I've had many clients who invested 10 percent of their projected gross sales only to have sales rise to a point where the absolute dollars represented by the 10 percent soon became a mere 2 percent of total sales . . . then even less. That never would have happened had the clients not been the soul of aggression on the marketing front.

- Being aggressive means learning all the marketing weapons that can be used.
- Being aggressive means seeing to it that all those marketing weapons *are used*.
- Being aggressive means using the marketing weapons consistently and more effectively each year.
- Being aggressive means knowing deep in your heart that you are truly being aggressive in marketing, using more marketing weapons than ever before.
- Being aggressive means having your competitors think of you as aggressive, as gaining new customers on all fronts.
- Being aggressive means having someone in your company, maybe even you, constantly thinking about marketing.

The third personality trait of the successful guerrilla is imagination.

I'm not referring to the ability to paint or draw or make up stories. Instead, I'm referring to imagination as it relates to your marketing message, your research, your competitive intelligence, your placement of signs, your selection of media, your allocation of marketing funds, your ability to determine the proper mailing list.

I had a client who noticed from his customer research questionnaires that a disproportionate number of them drove Porsches. So he rented a list of Porsche owners in his community, mailed to them, and broke the bank. That's what I mean by imagination.

When some marketing people hear the word "imagination," they think it refers to a headline or a jingle. The imagination I mean refers to marketing ideas rather than single efforts. An example is the tourist shops in San Juan, Puerto Rico, which sell to passengers on cruise ships. The most imaginative of these shops post the names of in-port ships in their windows, then invite people inside to see if their ship's room is the lucky number — room numbers being posted from all in-port ships. If the passenger's room number is posted, the passenger wins a prize. If the room number is not posted, well, the passenger sees a souvenir here, a souvenir there, and pretty soon loads up on souvenirs. The imagination comes from the realization, on the part of the shop owner, that the tourist ship passenger can be made to feel special, singled out, and the only one who can win the prize in the shop.

When one of the enormous hamburger chains (Burger King) opened right on the main drag in Miami Beach, one of the other enormous hamburger chains (McDonald's) purchased a sign one hundred yards away to point to its location, just a block off the main drag. Imagination in sign location saved the franchise owner tens of thousands of dollars in rent. Interestingly, I have since seen Burger King pull the same stunt on McDonald's. I guess that shows that imagination is even more important than originality.

More important than originality

Be careful when you exercise your imagination. Some marketers try to be imaginative when selecting a toll-free phone number, choosing what they consider a memorable number, such as 800-CALL JIM. But a guerrilla would hold back on the imagination and instead rely on information proving that few people will actually remember 800-CALL JIM for longer than a few minutes. So to a guerrilla, an imaginative toll-free phone number would be 800-453-7938. A number like that would just have to be written down, leaving nothing to chance — and memory.

The fourth personality trait of the successful guerrilla is sensitivity.

You've got to be sensitive to 360 degrees of reality. First, successful marketing guerrillas I've met were extremely sensitive to their marketplace. Most, if not all, were ultrasensitive to competitive activity and plans. Some had the imagination to buy stock in their competitors' companies, just so they could receive the annual reports. That certainly demonstrates their sensitivity to the competition.

Of course, you've got to be sensitive to your customers, to your prospects. You've got to have a sixth sense of their needs, their wants, their expectations. You must also be sensitive to your own geographic area. I once saw an extremely successful marketing executive from New York fall flat on his face in Denver, so insensitive was he to the needs of the local community and the entire Denver market.

Guerrillas I have known who achieved fame and fortune, especially fortune, through marketing, were very sensitive to the times. They would run one kind of marketing during times of prosperity, low unemployment, low inflation. Their marketing would do a quick 180-degree turn during a recession, during high unemployment, during rampant inflation. Although they remained with the same marketing theme, they orchestrated it entirely differently because of their sensitivity to the times.

Avoid cleverness

I always urge clients to develop a sensitivity to cleverness. Then I ask them to look for it in their marketing, and eliminate every single trace of it. Cleverness is the enemy of guerrilla marketing. Cleverness makes people remember the marketing, not the message. It makes the commercial memorable, but not the product or service. Cleverness in my opinion is the most misguided use of marketing.

The purpose of marketing is to create a desire to buy, to sell products and services on a repeat basis. The purpose is not to make people laugh, gain applause, win awards, use special effects, or have Myrna poke Harold in the ribs and say, "Oh Harold, wasn't that clever?"

A sensitivity to cleverness will automatically make your marketing more effective because the majority of marketing people have this misbegotten belief that marketing *is supposed to be clever*. They automatically think that headlines should have puns, that copy should have rhymes, that jingles are the only responsible use of music, that photos of women should be sexy, and that special effects are the name of the game on television.

Guerrillas also must be sensitive to their public. In the past, the popular fiction spread among the marketing community was that the public is as smart as a typical twelve-year-old. But times have changed. *These days, the public is as smart as your mother.* And you know that she's no dummy. You can't pull the wool over your mother's eyes. She won't be won over by clever marketing machinations and a surplus of adjectives. Just remember that you've got to be sensitive to your public and that your public is a savvy group, sharper these days than ever.

The public is as smart as your mother

The fifth personality trait of the successful guerrilla is ego strength.

I do not mean that you must be egotistical, conceited, and self-centered. If you are, customers and employees learn to resent you. Instead, I mean you must believe in your product or service fervently enough not to be self-conscious about marketing, fervently enough to mail letters, create brochures, or run ads that are loaded with information. You need the ego strength to tell your whole story and not worry about the copy being too long.

I also mean another kind of ego strength: You develop a

marketing plan. You create a marketing campaign. You establish a marketing theme, then you start marketing to beat the band.

Who are the first people to become bored with your marketing? Your employees. Who are the second to become bored with it? Your co-workers and associates. Who are the third? Your mate and family. The fourth? Your best friends.

The worst possible advice

All these people who love you dearly give you the worst possible advice. They tell you they are bored with your marketing and it is time to change it. They say they are bored with the campaign, the theme, the look, the format. It takes a powerful ego to stand up to these people. It takes an ego constructed of granite to tell the people who love you the most to keep their opinions out of your marketing.

Are your customers bored with your marketing? No way.

Are your prospects bored with your marketing? Never.

You'd have to run your marketing for centuries to bore your prospects. They barely know you exist. It hurts to realize this, but they probably don't care enough about your marketing to ever become bored with it. So don't change your marketing once you've launched it. That takes a powerful ego, but this is the time for that power to assert itself. Hug your loved ones, but don't listen to them when they speak of how tired they are of your marketing.

The designated guerrilla

If you cannot adjust your personality to incorporate these traits, you'd better assign a designated guerrilla to be patient and aggressive for you, to keep up-to-the-moment on the latest marketing tactics, to capitalize on marketing opportunities without breaking the budget, to stick with your plan, to exploit the vulnerabilities of the competition, *to show you how to intensify your attack constantly.*

To repeat, the personality characteristics of successful guerrillas I have known — in *Fortune* 500 companies, in ten-person businesses, in one-person shops, in start-up operations, the ones who have hit it big by the yardstick of profitability — have been:

1. Patience
2. Aggressiveness
3. Imagination
4. Sensitivity
5. Ego strength

If those personality characteristics describe you, you're in great shape. If they don't, delegate the marketing function to someone in your organization who is the very essence of them.

The guerrilla marketing attack is not just an exercise. It's a real battle, and there are both winners and losers. Now you know the personality of the winners.

3
The Seven-Word Winning Credo

A CREDO IS a formula of belief. The seven-word credo in this chapter is exactly that. It is not a statement, but a set of words to serve as a formula for your success in marketing.

Guerrilla marketers must memorize these words. More important, they must believe in each word with a depth of feeling that makes each one second nature. They must trust each idea expressed by the word enough to allow it to guide their decisions, outfox their competition at every turn, provide solid ground in shaky times.

Seven invisible concepts
Each of the seven words represents a basic concept that is in plain sight, yet appears invisible to the majority of business owners in America — as proven by the paucity of their marketing, their infatuation with cleverness, and the depressingly high rate of their business failure. Lack of marketing skill is not always the culprit. But it deserves the blame in more cases than you may imagine.

Businesses get started by all kinds of people. They are bright and experienced in one or more facets of business, but that rarely includes marketing. So they market by the seat of their pants — and that's exactly where they land.

The process of marketing is both a challenge and a cinch for a guerrilla. Being aware of the 100 weapons available for the guerrilla marketing attack, and feeling deep in your bones the seven-word credo, you know your attack will succeed in establishing a pattern that constantly earns high profits for your company. The weapons and credo are enough to win any battle. Setbacks? Maybe. But you will achieve your business goals because you know how to win. The seven-word credo will teach you how.

In order to make marketing easy for yourself, you're going to have to learn the gut meaning of seven words, understand the behavior they call for, and actually *live by the credo*. That's not easy.

It's an all-out attack we're talking about: using as many of the marketing weapons as possible, developing or polishing all five of the personality characteristics, living by all seven words of the seven-word credo.

Based on my own experience with large and small businesses, I can confidently predict failure if you live by fewer than all seven ingredients of the formula. Some of you may slip through the cracks and achieve flash-in-the-pan profits. But most of you will lose money and sleep if you try to slide by with only six words. I also predict success and profits if you incorporate the seven-word credo deep into the essence of your marketing. The longer you live by the credo, the less effort it takes. Soon, it takes no effort at all. It will be part of you.

Let me give you a memory crutch for the seven words. They all end in the letters "ent": *commitment, investment, consistent, confident, patient, assortment*, and *subsequent*. Now you have the memory crutch and now you have the credo.

Remember "ent"

1. COMMITMENT

Marketing does not work instantly. You know it takes nine impressions to motivate even a prospect to buy from you. And you know it takes three attempts to make each one of those impressions. A large number of business people run their marketing for a few weeks, then stand back so as not to be stampeded by the customers. But nothing happens. So they change their marketing. Again nothing happens. The more they change it, the more nothing happens. Eventually, they stop believing in marketing. Eventually, they are out of business.

An equally large number of business owners embark on a marketing program, then decide to change one element of the

Resist change in marketing

program — the places in which they advertise, or their message, or their format, or their basic offer. Big mistake.

A recent study of two products, a $10 device and a $10,000 package, revealed that for both it took four months from the date of the first ad until measurable sales increases due to advertising were recorded. Four months. Think of how many business owners abandon their marketing program after three and a half months!

Three months of faith

I tell my own clients to expect no results for the first three months. Anything before that is lucky, gravy, pure serendipity. After three months, they can expect some positive effects from the marketing. And each month thereafter, those effects should be better than the month before, with rare exceptions.

My clients are requested to stay with the program for the first three months no matter what happens. Preferably, they will agree to stay with the program a full year. But three months is usually enough for marketing to begin to prove its worth.

At that point, a few people might mention that they saw your ad or heard your commercial, received your mailing or noticed your sign, picked up your brochure or read your circular. They may even say they heard your commercial although you're running no commercials. The point is that you'll begin — just begin — to see results.

The recent study I mentioned said it took four months for the marketing to work. And the project that showed it took nine marketing impressions to motivate a purchase indicated a 27-week, or nearly seven-month, wait. Whichever study or opinion you wish to subscribe to, you can plainly see that to get to the point where marketing takes hold requires *commitment*. Marketing is similar to marriage in that both will work with enough commitment; neither has much chance without it. I hate admitting this about my own profession, but *brilliant marketing without commitment doesn't work as well as mediocre marketing with commitment*. It takes a barrel or more of commitment to stick in there and wait and wait while your prospects are noticing your marketing for the fifth time, the sixth time, the seventh time — and they're *still* not coming in to buy. What's going to keep you going? Commitment is.

When mediocre marketing works

Imagine being involved in a shipwreck. There you are, out at sea, far from land. Yet you can see land, so you decide to swim for the shore. We both know that you'll never make it without commitment. If you decide to swim for an hour or so, see what happens, then wait — you're sunk. The only way you're going to survive is with commitment. That's how marketing is akin to survival at sea. Keep swimming; you'll make it.

So it is with marketing. Develop a marketing plan, taking as long as you like to do it. But once you've done it, commit to it. The longer the time you commit, the better it will work for you. The shorter the time you commit, the worse it will work for you. Expect miracles and none will come. Commit to your plan and you'll get not miracles, but profits. Profits are the fruit of commitment.

2. INVESTMENT

Money that is put into marketing is an expense, to be sure. But more than that, it is an investment. It's only a conservative investment, but if you follow the seven-word credo, it's an investment that will almost always be more lucrative than stocks, certificates of deposit, treasury bills, tax-free municipal bonds, options, commodities, or real estate.

Forget, for the moment at least, the certainty and the size of the payoff of your marketing investment. Consider first the idea of marketing as an investment more than an expense. If you bought IBM stock regularly each month, you wouldn't consider that to be one of your expenses, but one of your investments. If you had in your possession a certificate good for a hundred shares of IBM stock, you'd take good care of that investment. If IBM stock dropped a couple of points, you'd never tear that certificate into shreds. That would be utter **Utter nonsense** nonsense! But utter nonsense goes on every day in the world **in marketing** of marketing. Business owners invest in marketing, don't achieve the results they hoped for within the time they hoped for, then discard their entire marketing plan, deciding to start again and go about things differently. That's just like ripping up that IBM stock certificate! The IBM stock will probably go up in a short while, enabling you to enjoy a profit on your investment. But you won't receive any profit if you turn the certificate into confetti. And if you abandon your marketing plan before it has a chance to work for you, you're depriving yourself of any hope for a return on your investment. You are taking your investment and flushing it down the toilet. People never flush their IBM investments down the toilet. But they flush away their marketing investments every day.

Why do they do such a ridiculous thing? It's because they never learned to think of marketing as an investment. They think of it as an expense. They never learned how commitment makes marketing pay off. They think that marketing works instantly, that human behavior is changed instantly. If you treat marketing as an investment, you'll be prone to spend more for it, to stay with it longer, to give it a chance to pay off, to expect profits instead of miracles. You won't be so quick to throw it away as ineffective before it has had a chance to do what marketing does.

Guerrillas in Marlboro country

Here's an example to show exactly what kind of return a marketing investment is capable of generating. Shortly after the introduction of Marlboro cigarettes, a research study indicated that they were perceived as having a feminine identity. It was surprising that this brand, one of the worst-selling in America, had any identity at all. So Marlboro invested in their now-famous marketing campaign centered on cowboys. Do you know what the effects of that marketing were after *one full year*?

There were hardly any effects! The brand still had a feminine identity. Sales were almost as sluggish as before the marketing started. But the Marlboro management people were committed to their campaign. They realized the money that they put into it was an investment.

Soon, Marlboro's identity began to change. More and more men bought Marlboros; more and more women bought them, too. The brand started climbing in sales to the point where it became America's number-one brand. Each year, Marlboro pulls even further ahead of the second-place brand. Yet Marlboro marketing is almost exactly as it was when the cowboy made his debut. The Marlboro people clearly understood the meaning of commitment, clearly recognized that their marketing was an investment. It paid off better than any other investment they could have made — *any* investment.

3. CONSISTENT

Earlier, I alerted you to the need for ego strength, one of the five personality traits of victorious guerrillas. You need it to stand up to those who will advise you to change your ad, your format, your theme, your media, your identity, your entire marketing campaign. I urged you to give those folks a cheerful wave, then send them scurrying.

Those people couldn't care less about your marketing being consistent. They know you. They know your company. They know your products or services. Regardless of what you did, you'd still have earned their faith. Most likely, they've known you for quite some time, or, if not, they've grown to know you well over a short period of time. But it's not going to be like that with your prospects.

They don't know you, your company, your products, your services, or anything about you — except for what they learn from your marketing. And they've got more important things to do than pay a heap of attention to your marketing.

To win their patronage, you're going to have to prove yourself **Prove yourself** stable, successful, sure of yourself. Believe me, you won't prove those things if you constantly change your marketing, your media, and your message. Rather than reassuring them, you'll be confusing them. But you can clarify your position in their minds rather than confuse it if you maintain a posture and stick with it. The idea is for you to remain consistent. Be consistent in your theme, your format, your graphics, your media, your identity.

You can change your ads. You can change your offers. You can change your headlines, your copy, even your visuals. But keep all those changes within the parameters of a consistent

Something to lean on

format. Give your prospects something they can lean on, something they can count on. Sway not in the wind. Stand tall in the face of time. Your prospects will feel comfort, security, safety in the consistency of your message.

Naturally, they will not think those thoughts consciously. But in the next chapter, you'll realize that it is unimportant what they think consciously. You will be making inroads into their deeper, inner minds. That's where the real action is. That's where the decision to spend money comes from.

Where the real action is

The moment you are inconsistent in your marketing is the moment when you shake their faith in you. They figure that if you're not sure about what you want to say or be, how can they be sure? Why should they risk their hard-earned dough on a potential fly-by-night? They shouldn't.

If there's anything they want from a new venture, it's the assurance that they won't go wrong if they buy from you. The more you remain consistent with your marketing, the more assured they will be. The more you change, the less assured they will be. I don't have to belabor that point.

True creativity in marketing

True creativity in marketing requires that you maintain a fresh feeling within your current marketing format. Anybody can make a radical change if they're willing to sacrifice consistency. But it takes a genius to make radical changes while staying consistent. For a list of many companies who have accomplished that, take a long look at the *Fortune* 500.

It's not that these superconglomerates are the essence of consistency, but many of them have maintained the same marketing posture for years — with profit and joy to show for it. Just remember, as a practicing guerrilla, you are not supposed to get your creative jollies from new art and copy, but from new customers and new highs in profits. As one ad agency justifiably argues, "It's not creative unless it sells." And I add, "It's going to sell a whole lot easier if you're consistent in your selling of it."

Be sure you know what to be consistent about. A major antacid advertiser was clearly determined to be consistent in the humor present in all its TV commercials. Now that's something no self-respecting guerrilla would hang consistency on.

Everyone remembered the funny commercials, even the punch lines. But they didn't remember the product. Did they remember the main benefit? No. Did they *buy the product?* No. Yet they did remember the humor. But when your tummy hurts, you don't need a laugh, do you?

4. CONFIDENT

Hardly anybody patronizes a business by accident. Money is parted with only after careful deliberation in an off-limits section of the brain as yet undiscovered by anyone.

Before people spend money, they want to be certain they're not wasting it. They look for indications that their money will be wisely spent. You can probably go to the bank if you know what most people consider "wisely spent."

I cannot tell you exactly what *your own prospects* are seeking in your specific business, though you'd better be quick to learn — by means of a questionnaire, direct conversation, or research **A question** of some kind. The question is certainly worth asking: What **worth asking** influences people to patronize businesses such as yours?

Research in the furniture industry shows the four most important things that influence people in choosing one store over another. I realize that your business may not be akin to this particular industry, but the lessons will still apply. Let's start with the fourth:

Number 4: People are looking for a business that offers a wide selection. They don't want to pick from one or two colors, styles, price ranges. They want a whole lot of choices. The more choices the better. And they'll take their money to a business that offers a broad selection. Of the countless factors that influence sales, selection comes in fourth. That means it is extremely important.

Number 3: People are looking for a business that provides excellent service. These days, people demand service, expect service, will go out of their way for service, will pay for service, and will select stores based on their reputation for rendering service. There are many determinants that govern a person's choice of stores. Service is the third most important.

Number 2: People are looking for a business that offers quality. I've been consistent in my reminders that even guerrilla marketing will not help sell shoddy quality more than once. You might have guessed that quality would be the number-one influence on sales, and you'd be close. Quality is worth paying extra for, worth traveling further for, worth waiting longer for. You might be interested to know that price came in number nine in the research. Everyone is out to get a good value. But they recognize that they may have to pay a bit more to gain that value. And there are factors more important than value. Quality alone is almost enough to win customers. It stands up there ahead of every single other factor, except for one.

Quality alone can win customers

Number 1: People are looking for a business in which they can be confident. That fact alone merits a place for the word "confident" in the seven-word credo. Commit the word to memory, then do your utmost to earn it.

Confidence is one of the goals of the guerrilla marketing attack. If people see that you are committed to your marketing, your media, your look, feel, tone, personality, attitude, position, market niche, and audience, they will feel more confident in your business.

If prospects see that your marketing appears regularly, that you are never out of the public eye, that you stand up to be counted week in and week out, and that there is consistency in your marketing, your presence, your tone, and the benefits you offer the world, they'll be more confident in your business.

The first three words of the credo, "commitment," "investment," and "consistent," lead to the fourth as they make your prospects more confident.

5. PATIENT

You already know that patience is a personality trait of the successful guerrilla marketer. I was unable to resist including it again in the credo. I hope you won't be able to resist including it either. It takes a patient person to practice commitment, treat marketing as an investment, keep marketing consistent, and stay on track until prospects have been made confident in the business. It's a good idea also to think of marketing as farming or gardening. First you prepare the soil. Then you plant the seeds. Then you fertilize. Then you water. And then you wait. And wait and wait and wait. *Finally,* the crop comes up for you.

The importance of patience

6. ASSORTMENT

You have been armed with a large assortment of 100 marketing weapons available to a guerrilla for the attack. By now, I hope you have a secret feeling of superiority over your competitors who are unaware of them. For your sake, I hope they never become aware of more than a handful of the weapons. If they use only six weapons, and you have an arsenal of 46 that you are using with aggressiveness, you've got a lot of ways to win the battle for the prospect. Your marketing inventory should contain the widest possible assortment of marketing weapons. In most cases, the bigger the assortment, the higher the profits and the happier the business owner.

Wide is wonderful

Here I go again, but you should know that your assortment of weapons doesn't mean more than an assortment of beans unless you are committed to using them, unless you understand that their cost is an investment, unless you are consistent in your handling of them, and unless you are patient enough to make them part of the credo that guides your marketing.

It would be simplistic to say that the larger your assortment of properly used marketing weapons, the larger your profits. So I won't. But pretend I did, and market according to that key principle.

7. SUBSEQUENT

The opposite of a guerrilla marketer is a person who thinks the marketing process is over when the sale is made. A card-carrying guerrilla knows that's just the beginning. The process starts when you conceive of your product or service and continues until you have the blessed patronage of repeat customers who keep sending friends and acquaintances in to buy what you're selling. It has been estimated that 80 percent of lost business is lost solely because of apathy after the sale.

Moral to guerrillas: Market like crazy subsequent to the sale. Realize that the love affair has just begun. Unlike the love 'em and leave 'em marketers who think the honeymoon is over when the sale is over, indelibly stamp in your mind that *the honeymoon is just beginning.*

The honeymoon should never end

Write or call your new customer thanking them for the sale. How many such thank-you notes or calls have *you* received? Hardly any. That proves most competitors are asleep at the cash register. And after you write or call your new customers, write or call them again, making them a new offer, suggesting a companion to their recent purchase. Later, write or call them again, gently suggesting that they refer your business to friends who might appreciate the selection, service, and quality that made them so confident in your offering. Later, write or call them again, making a special offer — possibly one you're extending only to them because they're so special to you.

Keep marketing to your customers subsequent to the sale. Friend, we're talking living, breathing *customers* here. Treat them like your long-lost brother — or better. Customers are a very rare breed. The vast majority of the people on planet Earth are not now and never will be your customers. I hate telling you that, but if you don't face up to it, you'll run the risk of neglecting that precious, lovable, endangered species — your customers. Make no mistake: They are endangered because new competitors every day are trying to woo them, if even for just one sale.

A precious, lovable, endangered species

The guerrilla marketing attack is a sustained attack. Once started, you never stop attacking. You may vary your weapons, adding some, subtracting others, polishing others. You may be guilty of a host of sins, but never be guilty of apathy toward customers subsequent to the sale.

• • •

"Commitment. Investment. Consistent. Confident. Patient. Assortment. Subsequent." There's your seven-word credo. Memorize it. Live by it. Profit by it.

As a sort of postscript to the credo, I offer you one more word, this one beginning *and* ending in "ent." The word describes what profits should be to you: your *enticement*.

4
From an Art to a Science

DURING THE EARLY DAYS of marketing, which probably began when a farmer asked the local town crier to mention that he had some hogs for sale, marketing was primarily an art, though the unspoken truth is that it was a business. It pained many creative types to admit they were businessmen or businesswomen rather than the artists they viewed themselves as. They reveled in the idea that they were involved in an art. Today, the spoken truth is that marketing is a business. But many creative types still wince when they must own up to that truism.

The transformation of marketing   Clearly, a major change has taken place. Marketing, that veteran art form, *is being transformed into a science.* That means less is left to chance; less is left to the whim of the word- and picture lovers; more is left to the tough-minded business-person who now wears the hat of the scientist.

As a guerrilla mounting an attack, be appreciative of the growing role of science in marketing. You and I have the psychologists of the world to thank for removing the guesswork from our efforts. Psychology intermingles so closely with marketing that many marketing firms now hire psychologists. I'm talking about psychology more than market research. Psychology is the study of human behavior. It stands to reason that the more you know about how humans behave, the more you'll be able to influence their purchase decisions.

When I was in college, majoring in psychology back in the prehistory of the early fifties, I well remember the words of the professor who told the class that there were no rules in psychology yet. It's still all theory, he told us. He said that people were trying to prove some of those theories, and someday they might. But at that time in the history of psychology, there were no laws.

A great deal of time has passed since the professor uttered that depressing thought. Some of the theories of human behavior we were studying were debunked. Happily, some of the other theories were proven correct. Increasing numbers of hypotheses have resulted in absolute laws. Psychology does have laws these days. And guerrilla marketers can derive deep and warm satisfaction from that fact.

They can uncross their fingers, knowing that many of their marketing efforts *will* pay off rather than *might* pay off. Art is gorgeous. But science is more precise. We guerrilla marketers can doff our caps to the psychologists who have gladly shared with the marketing world the fruits of their findings. Guerrillas feast hungrily on those fruits. They apply them willingly to their craft. They are all for art, but when it is enhanced with science, they breathe sighs of relief, and their eyes brighten with optimism. As psychology progresses, bolstered by other sciences, such as biochemistry, that add insights to our knowledge of human behavior, marketing will be more and more enriched.

Art belongs in galleries

Understand here and now that guerrilla marketing uses as much science as possible. Psychology is expanding in many directions. So let's focus on a few fundamental truths based on proven psychological theories that you might consider embroidering, framing, and hanging in your office.

Scientific fact: Most purchase decisions are made in the unconscious.

Great numbers of people see a commercial, but don't rush out the next day to buy the product. Large groups of people see an ad, but don't act on it as quickly as you'd like. When they finally do act on it, they may not even know why.

When I worked at a Chicago advertising agency, a group of four of us agency types flew to New York to meet with our client. We were working on a big new marketing campaign for a major corporation. We all felt excited about it and were eager to show it to our client.

En route from JFK Airport in the taxi, we talked about the upcoming meeting. The driver, overhearing our conversation, leaned back and asked, "You guys in the ad game?"

"Yes," we told him.

"You guys really believe that crap works?" he asked, and he earnestly wanted an answer.

"Sure we do," one of us said. "We wouldn't be having this meeting if we didn't."

The cab driver set us straight: "Well, I sure as hell don't believe it works. I can tell you this — I've never bought anything because of advertising and I never will."

"What kind of toothpaste do you use?" one of us asked.

"Well, I brush with Gleem," he told us. "But it's in no way connected with the *advertising*. It's because I drive a cab and I *really* can't brush after every meal."

That's a funny story, and true, based on Gleem's marketing theme at the time: "For people who can't brush after every meal." But it's not that funny when you consider how many products *you've* bought because of some aspect of the marketing. You won't be able to identify many. But if you go through your refrigerator, food cabinets, clothes closets, and medicine chest, and think of your appliances, car, and whatnot, you'll be a bit shocked at how much marketing has motivated you. You're not unique in this respect: It's the same way with almost everyone. That's why so much time passes before most marketing works. Accessing the unconscious mind of a human being is not all that easy.

Scientific fact: We now know how to access the unconscious mind of people. Repetition does the job.

Repetition does its job surely and reliably, but it doesn't do it quickly. It's true that fat-cat marketers can afford enough marketing exposure to repeat their message fifty times in one week. I once wrote a TV commercial for a Sears anniversary that ran fifty times *in one weekend* in most of the major markets in America. They were able to repeat their message and get into the unconscious minds of millions in a relatively short time.

There's not going to be such instant gratification for you. To get into the unconscious minds of your prospects fifty different times might take a hundred and fifty weeks. Or it might take a hundred and fifty different telephone calls. Or maybe a hundred and fifty different mailings.

Without question, the closest contact in marketing is pure face-to-face contact. Even with that level of relating, here's a rule of thumb for how many sales calls must be made before a sale can be consummated:

2 percent of sales closed:	One sales call
3 percent of sales closed:	Two sales calls
4 percent of sales closed:	Three sales calls
10 percent of sales closed:	Four sales calls
81 percent of sales closed:	Five or more sales calls

Those figures are for face-to-face marketing. They do not require accessing the unconscious, although that process certainly takes place during each call. In real life, you probably won't be able to afford that many sales calls. Industry research shows that in-person sales calls cost $275 each, including the salesperson's time and support materials; it shows 5.5 calls to close one sale. That's a huge investment per sale — over $1500 per closed sale. Unless you can afford that, and it might be an excellent idea depending upon your profit per sale, you've got

$275 per sales call

to sell through standard channels of distribution — places where people can go. And if you don't have those, direct marketing will be your prime marketing force. (Direct marketing for guerrillas will be fully discussed in Chapter 9.)

The more you have presold your prospects with other marketing, direct and otherwise, the more likely they are to buy your product. That preselling, even when prospects are not doing any buying, is conditioning them to buy. And that's happening in their unconscious. Are they aware of what's going on? Probably not. They have many more important things to do than to track your marketing. But you don't have many more important things to do than to influence your prospects on the unconscious level by repeating and repeating your basic offer to them.

The power of repetition

Repetition is of invaluable aid to you in many instances:

• Repeat your offer in each marketing message.
• Repeat your marketing to your prospects.
• Repeat your sales training to your salespeople.
• Repeat your marketing goals to your employees.
• Repeat your business goals to yourself.

Scientific fact: Your marketing can be twice as effective if you aim it at both right-brained people and left-brained people.

During my advertising agency days when my well-heeled clients applied more cash than science to marketing, I would have been guffawed out of many a conference room if I so much as mentioned right-brained people and left-brained people. Psychologists have since proved to us that half the people in America are left-brained, influenced by logical, sequential reasoning. And half the people are right-brained, influenced by emotional, aesthetic appeals. The real figures are 45 percent left-brained, 45 percent right-brained, and 10 percent balanced-brained. Like good guerrillas, let's go where the big numbers are and concentrate on the 90 percent left- and right-brained majority.

Half the marketing that goes on in this country misses its mark! The marketing in America that is directed to left-brained people — with powerful arguments, logically set down and sequentially reasoned — goes right past the half of America that couldn't care less for those appeals. If there's no emotion, no mood, no artistic panache, the marketing scores a fat zero *with half of its audience!*

Half of marketing is ineffective

The marketing that is aimed at right-brained folks — with a dazzling display of communications, emotionally charged appeals, and a potent tug on the heartstrings — sails on by the half of Americans who are absolutely not turned on by those kinds of entreaties.

Does that mean that approximately 50 percent of money spent on marketing in America is wasted? You bet your bank account it does!

Marketing that does not take cognizance of the brain orientation of its audience is irresponsible, behind the times, wasteful, and ill advised. Guerrilla marketing always honors the brain orientation of its prospects. Guerrilla marketing, employing this scientific finding with the precision of a surgeon, incorporates logical, sequential reasoning *plus* emotional, aesthetic appeals in all marketing messages. Guerrillas don't want to waste their money, don't want to miss their market. Guerrillas use science to gain customers. Their marketing talks to *all* their prospects.

Even some of the giants are employing the tactics of the guerrillas. You can see examples of such marketing in many of the efforts of Apple Computer, Rolls-Royce, and Mobil Oil.

Scientific fact: The more statistics you have, the more precise you can be in your marketing. And statistics of all kinds are now available through the U.S. Government Printing Office.

Just call them at (202) 738-3238, and request their latest edition of *The Statistical Abstract of The United States* ($25 paperback; $30 hardbound). This will provide you with crucial data about your prospects, information that can make them your customers because it allows you to be specific in your marketing. You'll find 1400 statistical tables that will enable you to use science to persuade your potential customers that they ought to be paying customers. Few businesses exist that don't base their important marketing decisions on some sort of statistical data.

Scientific fact: Children are influencing family purchases more today than ever before.

About 60 percent of children aged six to fourteen influence family purchases of such items as TVs, stereos, VCRs, and microwaves, and even help decide where the family goes on vacation. Mothers estimate the average value of purchases influenced by children to be $300 per year. Since 70 percent of today's mothers work part or full time, they are more lenient with their kids and will involve them in family purchasing decisions. The days of "children should be seen and not heard" have come to an end across most of America.

Scientific fact: It is now possible to predict the behavior of many consumers, but you can't rely on the consumers to provide accurate information.

You can predict consumer behavior if you are aware of people's attitudes — their honest attitudes, not their stated attitudes. Consider: 96 percent of all adults would like to change something about their appearance, but they don't really do much about it. And 76 percent claim they exercise regularly, yet 59 percent are overweight. More than 50 percent say they try to eat a healthy diet, but 38 percent who dine out order a meat main course, and 37 percent say they'd like to have more "all you can eat" restaurants available. This is not because people are dishonest. It's just that when the moment comes to make a buying decision, they don't always do what they say they'll do.

Scientific fact: People's values are now measured and ranked. Knowledge of them can dramatically aid your marketing. But be warned: These values change regularly.

In 1974, the top seven values were, in order, freedom, happiness, wisdom, self-respect, mature love, a sense of accomplishment, and true friendship. In 1988, they were a satisfying family life, good health, enjoyable work, peace of mind, close friends, wealth, and free time. Address your marketing to these values, and keep abreast of them.

Scientific fact: There are two bonds to make to speed the closing of a sale — the human bond and the business bond.

People would *much rather* do business with a friend than with anyone else. You can become their friend if you make a human bond with them before you make a business bond. The human bond can be as short as one or two sentences, with a comment on your part, then a response on the part of your prospect.

The human bond can be about anything — except business. Qualifying topics include the weather, the day, the family, sports, the news, gossip, facts, almost anything — except business. Spend a moment to establish this bond, whether you use personal calls, telemarketing, or any other face-to-face marketing. Smiles and eye contact do wonders to establish the bond, and mentioning the prospect by name is another powerful bonding method.

It's a lot more difficult to establish the human bond if your prospect is not physically there to respond. Still, if you make the try for a human-to-human connection, you'll get in a lot closer than if you ignore it entirely. Don't be phony. But be human.

After you make the human bond, it will be much easier to set up the business bond, and then the sale. The human bond is easy to reinforce when next you meet or communicate. For example, try a letter or phone call that starts out:

Get to know your prospects

Dear Jim,

I hope you enjoyed that fishing trip and caught your limit every day. I'm trying to give you that same kind of satisfaction with this special offer that . . .

You can't always make that human bond, but the more aware of its importance you are, the more bonds — and sales — you'll make. Is it time consuming to gain enough information to make the human bond? It sure is. And it's worth every second.

Scientific fact: One of the most important human needs is for an identity. Recognize the identity of your prospects.

Don't treat people as prospects. Don't treat them as members of a demographic group. Recognize their specialness, then treat them as the special individuals they are. For instance, recognize your prospect not only as a fellow human but as Nicole Denise Pope, who breeds tropical fish as a hobby. Mention Nicole's interests to her whenever you can. Find out more about them. When Nicole discovers that you treat her as the special person she knows herself to be, she'll know that you'll offer her the same care and sensitivity *after* she buys what you're selling. You will have increased her confidence in you. You will have established your own business as special by recognizing Nicole Denise Pope as special. Harvey Mackay, president of the Mackay Envelope Company of Minneapolis, sees to it that his company learns 66 things about each customer. He calls this policy "The Mackay 66" and it has worked well enough for his company to produce ten million envelopes per day.

Scientific fact: People have a basic need to belong. Let them belong to your "club."

Can you create a clubby feeling for your prospects and customers? Of course you can. Here are a few ways:

• Put them on your newsletter subscription list.

- Give them a membership certificate or card.
- Invite them to private sales.
- Provide them with advance and inside information.
- Greet them by name and use their name liberally.
- Send them a gift with your company name and their name on it.
- Mail to them frequently.
- Use warm and friendly wording in your communications.
- Think of what you can give them, not sell to them.

Just open your mind wide and think of other things you can do to intensify the feeling of belonging for your customers. They'll appreciate the attention; you'll appreciate the profits.

Scientific fact: People love to be recognized as experts.

That means you should tap the expertise of your customers and prospects. Provide them with questionnaires that solicit their opinions. If you're trying to sell something to them, ask their advice rather than trying to sell. They'll appreciate you for realizing that they are, indeed, experts. And they'll be free with their words of wisdom for you. Not only will you win a friend, but you'll also win a sale. And you'll increase their proclivity to engage in word-of-mouth marketing for you.

Ask yourself how many businesses have treated you as an expert. How would you feel about companies that did? Mightn't you be motivated to do business with companies that have shown they respect your mind? You can be sure that your customers will come up with the same answers to these questions as you will. And in many cases, they'll show their gratitude by buying from you and recommending your products or services or both.

Scientific fact: Getting a person to say yes to a sale works best if you establish momentum first with lesser questions to which it's easy to answer yes.

Rather than go for the big yes right off the bat, go for it in little steps, with little yeses leading the way. If you ask a customer, "Do you want to buy this computer from me?" there's small likelihood that you'll receive a positive answer right off the bat. But asking "Are you looking for a computer?" might earn a yes. Following that with "Are you aware of any ways a computer can help you?" is again likely to increase the yes momentum. Once you've established in your prospect an unconscious habit of saying yes, that big yes will be easier to obtain.

As marketing is a process rather than a single event, the obtaining of a yes is also a process. I have a client who asks in his yellow pages ads, "Want more TV channels?" You can be sure that almost anyone perusing the satellite television pages in a yellow pages directory will unconsciously answer yes to that question. Innocuous? Maybe. But the momentum has started.

Scientific fact: Your customers will be buying a lot more than merely your product or service.

I don't think I really have to tell you this; you must have realized it already. But because I want your guerrilla marketing attack to succeed, I want to drum it in. So bear with me as I tell you what your customers are really buying when they buy your product or service:

What customers really buy

• They are buying your personality.

- They are buying your reputation.
- They are buying your attire.
- They are buying your store or office decor.
- They are buying your service.
- They are buying the aroma of your business.
- They are buying your package.
- They are buying the status of your offering.
- They are buying the orderliness of your business.
- They are buying your acceptance by the community.

This last point deserves a bit of clarification. Unlike what you may have heard, people *do not* want to be the first on their block to buy anything. They detest the idea of being pioneers. They know pioneers get arrows in the back of the neck. After all, they might make a mistake patronizing your business. No customer wants to risk that embarrassment or misguided expense.

Your job   Your job is to convince them of your acceptance by the entire community or industry. Let them know they are not pioneering. Use phrases such as "accepted by your community," "proven right here in town," "guaranteed to satisfy you just as it has satisfied others like you," and other terms that assure them they are not risking the making of an error.

Only a tiny percentage of your prospects will want to be the first to buy what you are selling. The vast majority have a "wait and see" attitude. They might yearn for what you are selling. But they are not willing to go out on a limb as potential guinea pigs.

Scientific fact: People will remember the most fascinating part of your marketing and not necessarily the product or service that you are marketing.

That's why you've got to be careful every step of the way. If you use a celebrity spokesperson, people may remember the celebrity and not your offering. If you use humor, the same applies. Show a magnificent piece of art and show your product. Which do you think is more fascinating to your prospects? Don't expect people to be interested in you. Remember that they are interested in themselves. So if you want to interest them, relate your offering directly to them, and do it in such a way that your product or service is the most fascinating part of your message. Your marketing need not be fascinating — but your offering had better be ultrafascinating.

Good guerrillas know that there are two schools of marketing — Freudian, based on changing the prospect's attitude, and Skinnerian, based on modifying the prospect's behavior. Which school is better? Both schools are better. Guerrillas begin to change prospect attitudes with Freudian marketing, then go in for the kill with Skinnerian marketing. The combination of the two is incredibly potent. You'd do well to know both, practice both, and profit with both. You can learn the details in *Guerrilla Marketing.* **Two schools of marketing**

There's one more key point to understand. It is based on a multitude of exit interviews with purchasers of products immediately after they made their purchase. The point is that *you need a share of mind before you can earn a share of market.* There is an enormous difference between the two. **A share of mind**

A share of mind is a place in a prospect's unconscious, a place where your company name and benefits are known and trusted. A share of market is how much of the total spent in your industry is spent on your product or service. It's easy to see how the share of mind must precede the share of market. Gaining that share of mind requires a fair bit of psychology — and patience.

An equally key point to remember is that *you must sell yourself before you can sell your product or service.* You probably already know that. But I've seen too many instances where otherwise smart business people demonstrated that they did not know it, so I am reminding you of it one more time. **Sell yourself first**

There's a raft of competition out there. They may be selling the exact products and services as you. They may even offer them at lower prices. But there is one thing they are not and cannot sell: *your personality.* It has become a scientific fact that people buy that personality before they buy the product. If not, everyone would be driving the cheapest cars and wearing the cheapest clothes. After all, both do the job: providing transportation, covering the body, and providing weather protection. But people will buy what is called the "overvalues" of your offering — the unstated benefits, the status that goes along with the purchase.

Did you know that the average college graduate of today is better informed than the average fifty-year-old? This is true because the average fifty-year-olds completed their normal education several decades ago. Since that time, unless they have kept up with all the periodicals, they have missed out on a lot of crucial information.

If you're interested in keeping up with marketing's transformation into a science, I urge you to read the psychological journals, such as *Psychology Today.* They are several months, if not years, ahead of the marketing journals when it comes to reporting on human behavior and how new findings can improve your marketing. The more of that scientific research you use, the better off you'll be. But I hope you will respect the relationship between your own business instincts and scientific research.

A tool, not a master   *The science is there for you to use as a guide. It is only a tool; it is not a master.* I have let you in on these scientific findings to provide you with more advanced weaponry for your guerrilla marketing attack. If you're to succeed as a guerrilla, you should be aware of the scientific facts whether you plan to follow them or disregard them.

5
Why Launch an Attack?

THE OBVIOUS ADVANTAGE of going on the offensive is that it gives you more control than you have on the defense. When you are in an attack mode, your competitors have to react to you, rather than proceeding with their own plan. Instead, they are guided by *your* plan. Frequently, they will overreact, such as cutting their prices to match yours, thereby negatively affecting their profits. Or they will change creative strategies in the middle of the stream. Sometimes they will aim their marketing at fighting you — which only calls attention to your offering without necessitating any increase in your marketing expenses.

An unstated goal of the guerrilla marketing attack is cost efficiency. The goal is attained when your marketing is active and committed on all fronts. Bless your competitors for being inconsistent, uncohesive, spasmodic, unscientific, and armed with too small an arsenal. They are not likely to launch an attack. This is a great advantage to you — and a reason to take the offensive.

Companies that launch guerrilla marketing attacks often force their competition into making tactical blunders. And naturally, if you are launching a full-scale attack, you are taking the stance of a leader — thereby gaining crucial consumer confidence.

The stance of a leader

When your prospects see that you are going all out — which they assume when they are exposed to your multiple marketing weapons — they automatically and unconsciously also assume that you are proud of your offering, that you are taking on all comers, that you have something to crow about.

As I keep reminding you, your product or service had better be good, because the launching of a guerrilla marketing attack exposes poor quality faster than no attack at all. But if your offering lives up to your marketing claims, the attack can increase your share of market, not to mention beautifying your bottom line.

Just what does it take to launch the attack?

Although a guerrilla marketing attack has only four components, don't let that fool you into thinking it's a cinch to launch one. Realize that unlike other competitive events, where an attack is a temporary show of force and aggressiveness, a guerrilla marketing attack is a *permanent* show of force and aggressiveness. It knocks the competition off balance, then *keeps* them off balance — until they, too, launch a similar attack. But don't worry: Few of your competitors are guerrillas. They may know how to spend money, but they probably don't know beans about a full-scale, long-term guerrilla marketing attack.

The four components of a guerrilla marketing attack are:

Permanence is paramount

1. SELECTING YOUR WEAPONS

In Chapter 1 you learned of the profusion of weapons available to you. The more you select, the more effective your attack will be. Just remember that you should only select weapons that are appropriate to your business and that you can use properly. This does not mean dipping your toes in the water, but plunging in up to your neck. Don't get in over your head by spending too much money or by devoting too much of your time. Use as many weapons as intelligently possible to mount your attack.

2. PLANNING YOUR STRATEGY

All guerrilla marketing attacks begin with a clear, understand- **Begin with a**
able strategy, as I discussed in *Guerrilla Marketing*. In brief, **strategy**
I'll repeat the elements here.

Page One Bookstore is a typical retailer, located on a charm-
ing suburban street and owned by a woman with experience
in bookselling, but not in marketing. To fill her marketing
void, she read *Guerrilla Marketing* and developed her own
marketing strategy. In two years, her sales quadrupled. So I
reprint her strategy here, just to show the simplicity of a winning
plan. Like all guerrilla marketing strategies, hers has seven
simple parts:

A. A statement that tells the purpose of her marketing: "The
   purpose of Page One Bookstore marketing is to build an
   increasing base of repeat customers."
B. A statement of how she will accomplish her purpose through
   marketing, focusing on her key benefits: "This will be
   achieved by stressing the selection of book and nonbook
   items within the store."
C. A statement that describes her target audience: "Our target
   audience is book-buying adult females within a one-mile
   radius of Page One Bookstore."
D. A statement that lists the marketing weapons she will employ: "Marketing vehicles to be employed will include newspaper ads run weekly in three newspapers, a window display
   that changes weekly, numerous signs inside that merchandise and cross-merchandise, a yellow pages ad, quarterly
   autograph parties, quarterly author lectures, quarterly in-
   store seminars, FM radio advertising during peak book-
   buying periods, postcard mailings every two months, brochures, a catalog, a one-time magazine ad with an ample
   supply of reprints, a book-selling area at all local conferences, accessing all available co-op funds, utilizing display
   fixtures, putting up book posters, and using our marketing
   theme on bags, bookmarks, premium gifts, gift certificates,
   and sales receipts."

E. A statement that positions her in the market: "Page One's niche will be one that stands for careful selection tailored for the community."

F. A statement that spells out her identity: "Our identity will be portrayed as warm, honest, knowledgeable, up-to-date, and ultrafriendly as proven by our greeting people by name, taking phone orders, setting up charge accounts, shipping anywhere in the world, and doing free gift wrapping."

G. A statement that tells her marketing budget, expressed as a percentage of her projected gross sales: "Ten percent of projected gross sales will be devoted to marketing."

3. CREATING YOUR MARKETING CALENDAR

This twelve-month calendar lists the marketing activities in which you will be engaging *each week of the year*. That means your calendar will have 52 rows, one for each week. It will also have five columns. The first is simply the number of the week. The second tells the thrust of your marketing that week (sale, special offer, new merchandise, focus on service, whatever you select). The third tells which weapons you'll be employing that week. The fourth tells how much money you invested in marketing that week. And the fifth, *which you complete at the end of the week, by feeling, then again at the end of the year, by numbers*, rates the week on a 1-to-10 scale, as to traffic, sales, and profits. Because you can't relate the exact profits to the marketing, only a rating is required. At the end of the year, with all ratings in place, and going by gut judgment and arithmetic, you'll be able to plan your following year's calendar with more certainty — eliminating the tactics that resulted in poor weeks, heavying up on those tactics that resulted in excellent weeks. After three years of this type of analysis and action, you should be ready to enter marketing nirvana — development of a calendar comprising proven winners only.

The framework for such a calendar looks like this:

Week	Thrust	Weapons Used	Cost	Results
1	Reference Books	Newspaper, Postcards	$200	7
2	Author Lecture	Newspaper, Lecture	$150	9
3	Reference Books	Newspaper	$150	8
4	Children's Books	Three Newspapers	$350	8

More detailed examples of guerrilla marketing calendars for other types of businesses may be found in Chapter 6 of *Guerrilla Marketing*.

4. REMEMBERING YOUR SEVEN-WORD CREDO

Your guerrilla marketing attack will fizzle at the expense of your bank account, possibly even your entire business, if you don't live by the seven concepts described in Chapter 3. Your strategy and calendar are the brains of your attack. Your ability to live by the credo is the power and strength that will help assure victory for you.

Launching a guerrilla marketing attack — or, for that matter, engaging in any marketing with fewer than these four components — is like stringing beads without tying a knot on the end of the string.

Tie the knot

It doesn't take much time or effort to select as many marketing weapons as feasible from an arsenal of 100. Creating the guerrilla marketing strategy will probably take a bit longer because it's likely that you'll have to do some research. Development of the marketing calendar certainly won't take more than a day — even though you have 208 blanks to fill in at the outset and 52 blanks one year later—the results.

The temptation to make changes is always there. Guerrillas resist that temptation. They know that succumbing to it hurts their consistency and risks reducing consumer confidence. But it will take time before you see the tangible effects of your marketing. My mentor, Leo Burnett, founder and guiding spirit of the giant yet consistently excellent advertising agency bearing his name, once said, "I have learned that any fool can write a bad ad, but that it takes a real genius to keep his hands off a good one."

Sometimes it will probably be necessary to make some changes. That's okay. Flexibility is encouraged, but remember that the line between flexibility and commitment is a faint one. If you feel you must make changes of any kind, I hope they fall into the category of adjustments rather than overhauls. If you wonder where the line between flexibility and commitment can be found, it's in the *degree* of change. Adjustments do not affect the soul and spirit of a guerrilla marketing attack. Overhauls undermine the commitment that sustains the attack.

The soul and spirit

A hallmark of your commitment to your guerrilla marketing attack will be your patience during the early months. You'll have to exhibit that same steadfastness during the slow weeks that are certain to pop up occasionally even after those first months in the trenches. Guerrillas look at trends, not at individual moments. Although their calendar is precise enough to show them a week-by-week glimpse of the future, their patience is powerful enough to let them examine trends from month to month. Bad weeks don't bother them. Bad months do.

Look at trends

After the first three to six months of their attack, guerrillas expect every month to be more profitable than the month before. The frequent meeting of these high expectations outweighs the frustration of sometimes falling short of them. After the first year of their guerrilla marketing attack, guerrillas expect every month to be more profitable than the same month the year before. If they've been true guerrillas they are rarely disappointed. This accomplishment provides them with the impetus to continue the attack. Eventually, guerrillas are besieged by competitors mounting their own attacks. Fortunately, few of these competitors are guerrillas. And all of them have given the guerrilla a head start.

What kind of marketing is tailored to the reality of your budget?

The answer is: the guerrilla marketing attack. Although it is not advisable to try to market with no investment, the guerrilla marketing attack allows you to use an abundance of marketing weapons regardless of the size of your budget. Some guerrillas invest $300,000 per month. Others invest $300 per month. All are guerrillas if they think and act like guerrillas.

Unlike standard textbook marketing methods, which cost an arm and a leg — with discounts — the guerrilla marketing attack can be adapted to any size budget. So it is within the reach of new and small companies. After enough time, those new and small companies become older and larger companies. As they grow, their calendar grows with them, adding new media, marketing with greater frequency, employing proven promotions. You can be sure that they earn healthy profits because they continue to function as guerrillas in marketing — only this time with larger budgets and more ambitious calendars. Still, the size of a budget does not prohibit any business from engaging in fierce and aggressive guerrilla marketing.

Anyone can afford it

What kind of marketing lets you see into the future?

The answer is: the guerrilla marketing attack. By providing yourself with a calendar that projects a full year ahead, you have the equivalent of a crystal ball. This foresight enables you to plan inventories, budgets, personnel, and cash flow because it shows your future so clearly.

Because so many people are interested in instant gratification, guerrillas can use their calendars to be certain their stock is sufficient when they undertake a particular marketing task.

Gratify them instantly

There are few things more disappointing to a red-blooded capitalist than attracting the right people with the right marketing, then having them purchase from a competitor because you are out of stock at the time they come in. Guerrillas, with their uncanny ability to see into the future, rarely encounter such situations.

What kind of marketing prevents marketing emergencies?

The answer is: the guerrilla marketing attack. In many small companies, the question of the week is often, "What should we do about marketing this week?" Or "We've reserved the space in the newspaper but the ad isn't ready! What can we do?"

These kinds of emergencies, not minor if you've been through one, exist no more when you've launched a guerrilla marketing attack. You know exactly what you'll do about marketing every single week. You always have the ad ready and eliminate the last-minute rushes that impede the creative process. If many ads you see in the newspaper look as though they were put together at the last moment, it's because they were.

What kind of marketing aids decision making?

The answer is: the guerrilla marketing attack. Because of its ability to show you future needs, you can make many decisions with aplomb. Service decisions, expansion decisions, purchase decisions, personnel decisions, vacation decisions — these usually tough decisions become much less tough if you know what you'll be doing down the road.

The calendar you so wisely prepared gets the credit for making these decisions much easier than they normally are. Without the guerrilla calendar or the guerrilla strategy, deciding has a lot in common with groping in the dark. The calendar and plan shed enough light to give you a clear vision for many of your decisions, marketing or otherwise.

What kind of marketing gives you a cohesive identity and adds power to each marketing element that you use?

The answer is: the guerrilla marketing attack. In many cases, a company will hire one consultant to design a logo, another expert to write their direct-mail letters, another pro to create their ads, another to conduct their sales training, and still another adviser to handle their telemarketing. And let's not forget that other guy who designed their yellow pages ads.

The result is a hodgepodge of marketing, all pulling in different directions, all saying different things about the company, all portraying a different identity. The guerrilla marketing attack calls for all weapons to come from a single source so that all are aimed at the same target. The cohesion of identity in all **Gain some** marketing has a synergistic effect so that two and two equal **synergy** more than four.

When people see your direct mailings, they remember your ads. When they see your sign, they remember your direct mailings. When they receive your call, they remember your sign. The salespeople say exactly what the brochures say. Everything is working together. Everything is pulling in the same direction. The prospects feel reinforcement of their desire to buy from you in the first place.

Good marketing presses certain "hot buttons" in the minds **The hot buttons** of prospects. They get excited about some feature of a company's product or service, such as its speed or convenience or simplicity, because that feature was stressed in the marketing. But when they call or come in and talk to a salesperson, that salesperson stresses the economy or the innovativeness or the craftsmanship. Now, those may be product advantages. *But* *they are not the reason the prospect came in.* They do not press the "hot buttons." They result in lost sales, disillusioned prospects.

You can be sure that guerrillas see to it that their salespeople read every word of their marketing so that the salespeople can press the same "hot buttons" that brought in the prospects. The more cohesive your marketing and your identity, the more confidence your prospects will have in your offering. And I sure don't have to remind you about the importance of confidence, do I?

Guerrillas know that marketing is multifaceted, but not that difficult to understand. Instead of being put off by the process, they actually enjoy it. One of the main reasons they enjoy it is because they are in total control of it. They see marketing as a challenge, as a function in which they have the upper hand over their competition — even if their competition can outspend them.

Because guerrillas are not scared silly by marketing, they pursue it more aggressively. They are bolder in their use of the media, in the creative expressions of their marketing strategy. The guerrilla marketing attack is something that they are masterminding. They are the generals and they are on the attack. If there is to be victory, there can be no mystique to cloud the terrain. And there isn't.

The marketing mystique

Of the many advantages of launching a guerrilla marketing attack, the removal of the marketing mystique is probably the most important. Sure, it's great to engage in an attack that is programmed to succeed. But marketing is ever-changing. And if there is no mystique blurring the changes, each change is a new chance to gain a greater share of mind, and then the share of market that follows.

When you launch your guerrilla marketing attack, you will have examined all the options, selected the best for your needs, and gone all out for victory. Armed with your multitude of weapons, your guerrilla strategy, your guerrilla calendar, and your guerrilla mindset, you are a general who is destined for a victorious attack.

Sure, it would help to have megasums to invest in that victory. But you don't need megasums. Instead, you have the battle plan that can substitute for an enormous budget. And you are on the offensive with a well-planned, carefully conceived attack. As long as you maintain that attack, you will succeed.

6
Basic Training for Guerrillas

IN NUMEROUS CLIENT MEETINGS I've attended and seminars I've conducted, I've been quite enthusiastic about a marketing idea, only to be greeted by blank stares on the part of the people listening. That's my fault. That's because I assumed the people knew the fundamentals of marketing. They didn't.

Even today, as I observe most marketing, even by otherwise laudable members of the *Fortune* 500, I witness a vivid demonstration that not all the titans have a firm grasp of the fundamentals. They can't really define marketing, don't know what it really means, view it too narrowly, avoid making decisions about it, and tend to copy their competitors rather than innovate.

Who can blame them? Marketing is shrouded in mystery and complexity for many bright people. They can talk for two hours about quantum physics or Japanese mass production techniques — but when they talk to you about marketing, their eyes glaze over.

Marketing is supposed to create a desire to buy products and services. Yet you can spot legions of marketing examples that **The truth made** create absolutely no desire to buy anything. Marketing is simply **interesting** the truth made interesting and communicated to all interested parties. But it is perceived by many otherwise sane people as some kind of smoky and mysterious wizardry. What's going wrong?

What's happening is that many people involved in the marketing process fall into the same trap that ensnared me: They assume that people know more than they really do. People do know a great deal — but few know a great deal about marketing, especially the basics. Texts and courses addressed to the MBA candidates are woefully weak in these basics. They have the advanced marketing techniques down pat. But you can't make it to the penthouse if you don't know how to get to the second floor.

Probably the most glaring example of giant firms failing in the basics is to be found in Silicon Valley, where bright engineering types spent fortunes hawking their new technology. But because some members of the computer industry overlooked one of the cardinal marketing rules — *be clear with your message* — their marketing was understood only by their competitors, their employees, and themselves.

The Silicon Valley failure

Much computer marketing certainly wasn't — and still isn't — clear to the vast majority of computer prospects who have not yet purchased a computer. It's clear only to computer experts. And most computer prospects are not computer experts. Judging by their marketing, computer marketing executives knew a heck of a lot about computers, but very little about human beings. Their marketing talked in bits and bytes, technobabble that was a language foreign to the majority of would-be computer owners.

As a result, millions of dollars have been wasted because computer marketing was aimed over the heads of prime prospects, and ended up being understood only by other members of the computer industry who already owned a computer and weren't in the market for another quite yet.

I've been driving my car for over 300,000 miles. I'm really not sure what makes it go. I am sure that I don't care, just as long as it keeps going. Marketing that talked about pistons and cylinder heads and compression ratios would not have attracted me to my car. I was attracted by plain English. But plain English was a lost language in my personal quest for a computer. I sent away to twelve computer companies, requesting brochures. Of the twelve, eleven were so technical that they scared me, bored me, and prevented me from making it past the first few paragraphs.

Fortunately, the twelfth brochure, created by Epson, was in

clear English. I could understand the words, the sentences, the paragraphs, the benefits, and the features of the computer. Of course I bought one. After five years of ownership, I still don't know about its bits or bytes, and I don't want to know.

I was able to understand the manual, able to get right to work with the computer, and able to use it regularly without having to take one class, course, or tutoring session. For the sake of you non–computer owners — the whopping majority of the American population — if you're about to buy a computer, be warned: If you can't understand the marketing, you won't understand the user manual. And if you don't understand the manual, you won't use the computer properly — if at all.

I am still floored when I think that such a large and sophisticated industry could have spent so much on marketing that was unclear and incomprehensible — even to people who wanted to buy a computer.

In addition to the guerrilla basic training rules that follow, make it second nature that all of your marketing be clearly understood by your prospects. If only 10 percent of your audience is hazy about your message, and you market to a million people, that means one hundred thousand of them won't know what you're talking about.

A few words about your image: Get rid of it.

Companies that decide to communicate a certain image in their marketing are generally companies that set out to mislead the public intentionally. And the public, which does not take lightly to misrepresentation, shows its displeasure by staying away in droves from the products and services offered by the companies with the image.

An image implies something false. It suggests a façade. It is the very spirit of phoniness. When a company markets its image and the public sees that the company is not exactly what it represents itself to be, the public feels ripped off.

This doesn't happen if you communicate not your image, but your *identity*. An identity is honest. It is not phony. It does not mislead. When a company decides to communicate its honest identity in marketing, the public sees that the company is exactly what it holds itself up to be. This breeds confidence. And confidence adds verve to the sales curve. **Identity instead of image**

If you're not happy with your identity, change it. Don't lie about it. Once you are satisfied with your identity, be proud of it. Herald it to the skies. This is not to say that I am all in favor of marketing your identity only. But it is to say that you can and should convey your identity in all of your marketing. That identity is your personality, and your personality can win friends and influence people, not to mention sales and profits. A personality helps you sell yourself, and if you don't sell yourself, you'll have a heck of a time selling whatever it is that you sell.

The key is not to decide whether to engage in identity or traditional marketing; the key is to blend both. In December 1987, *Forbes* magazine said, "The trend now seems to be moving . . . to blend powerful image ads with the more traditional form, which relied on specific selling points to hammer home particular and perhaps even unique product benefits." Although they used the wrong I-word — image — instead of the right I-word — identity — I do agree with their assessment of the current marketing scene.

As companies have identities, so do individual products. The great Procter & Gamble, probably the world's most sophisticated and successful marketer, used to advertise their fabric softener, Bounce, with the theme line, "Bounce is for clothes you can't wait to jump into." Cute, but come on now, P&G, get serious. This is for money. Several months later, the theme evolved to a more sensible "Bounce gives clothes softness with no static cling." The copy in the TV commercial used this theme line to communicate the nuts and bolts; the visuals, puffy white clouds in a bright and beautiful outdoors location, conveyed the identity. That's the combination that produces profits — logical reasoning blended with a pleasant personality. In this case, Bounce, the product, conveyed the identity. The manufacturer, Procter & Gamble, conveyed zero identity.

Institutional advertising

Pure identity marketing, known also as institutional advertising and corporate image advertising, will always continue, since there just isn't a plethora of rational points to shout about when it comes to perfume, beer, and other birds-of-a-feather products. Identity marketing is also a boon to many car manufacturers, who need *something* to talk about, since their engineering people often give them very little.

As a guide, consider your product or service, your market, and this fact: According to SRI International of Menlo Park, California, 70 percent of the U.S. population is "outer directed" — meaning people who respond to marketing that will make them look better. Only 30 percent are "inner directed" — people who make buying decisions based on what the product will do for them rather than what it makes them look like. Inner-directed people buy things because of specific product benefits that are communicated in the marketing: longer warranties, better price, more flexibility, better performance. Sure, you can show them pretty pictures and play lovely music. But be sure you back those tugs on the heartstrings with hard facts that make them want to buy what you are selling. Which is *your* market? The answer gives lethal ammunition to whoever will write the copy for your marketing.

To whom do I market, and what do I say to them?

There's an easy answer to this one, an answer that is part of the basic training for any guerrilla. The answer is: Market to *someone* and say *something* to them.

Sounds pretty fundamental, huh? For a fundamental marketing truth, you'd be amazed at how many companies don't know it. Instead, some try in their marketing to speak to *everyone*. Or they try to tell them *everything*. The net result, naturally, is that they end up either saying *nothing* to *everyone*, or *everything* to *no one*.

The idea is to pick out your market — your someone — then decide the key points — your something — to communicate. Saying something to someone is a realistic and attainable goal. Saying more than that to more people than that can eat up enormous chunks of your valuable marketing budget.

Probably the saddest example of this comes again from the Valley of Silicon, where the computer industry said something to someone. They picked the right something — computers, — but the wrong someone — the general business population. They should have narrowed their sights or clarified their message. Had they done this, I, for one, would have enjoyed their product five years before I ended up doing so.

Talk more about your prospect than your product.

As Leo Burnett drilled into the heads of those who wrote advertising copy at his agency, "Don't tell people how good you make the goods; tell them how good your goods make them." In other words, think in terms of what your reader or listener is thinking, not in terms of what you are thinking. Once you've got your prospect intrigued with the possibilities of what your product can do, *then* you can talk about your product.

The standard wisdom is to market your features and benefits. A wiser wisdom suggests marketing your benefits and features. Have you been led to believe that it is crucial that you interest people in your marketing? Well, it is, but people won't pay a whit of attention to marketing. They'll only pay attention to what interests them, sincerely interests them. So you've got to catch their interest.

But how? Let me discuss now a subject in which all red-blooded human beings are and will always be automatically interested — *themselves.* Talk to your prospects and customers about themselves and you've got their rapt attention. Tell them what your product or service will do to improve their lives,

The subject that interests everyone

loves, wealth, health, career, looks, security, or leisure, and you're well on the way to making a sale to a grateful customer.

They don't want to buy your high-tech widget; they want to buy a better or easier life. The guerrilla marketing attack is focused on one person at a time, not on demographic groups. Even though you may start your attack by homing in on your target audience, be sure the attack is aimed at individuals, one by one. Prove that you think of them one by one with every word and picture in your marketing.

Don't walk away from an investment.

If your company has engaged in any past marketing and you decide to change your direction, that's fine. Just be sure you don't ignore your past marketing as though it never happened.

It did happen. It was an investment in becoming known, noticed, recognized, separated from the ranks of total strangers. In your new marketing, use your old marketing as a springboard, not as a crazy relative you refuse to acknowledge.

Use your springboard

Guerrillas have been known to develop new marketing strategies, themes, thrusts, identities, and formats. But no self-respecting guerrilla has ever walked clear away from a prior investment of dollars.

Your old marketing can help you utilize and redirect existing momentum. And you already know the fundamental truth that momentum is part of the process of marketing. If you've got it, use it. Don't waste it. You've made an investment. You've established momentum. Now you want to change your marketing direction. Do it. But save time and money by starting from where you left off, not from some new point in your prospect's consciousness.

A business owner I know wanted to switch from one medium, newspapers, to inserts. For weeks he mulled over the smoothest way to make the transition without losing the share of mind he had established. His solution was to reprint the best of his old ads in his four-page insert. Based upon his sales and profits resulting from his switch, the idea worked wonders to get him off to a running start with no loss of momentum.

A guerrilla has three markets and one sacred obligation.

As a guerrilla, you will always have three markets — the universe, your prospects, and your customers. You have the sacred obligation of initiating and maintaining a process of converting members of the universe into prospects, converting prospects into customers, and marketing with consistency and enthusiasm to your customers.

YOUR FIRST GUERRILLA MARKET: THE UNIVERSE

The first guerrilla market — and the largest of the three, yet the least profitable for you — is every single person in your marketing area. That might mean one neighborhood, one industry, one city, one nation, or even the entire universe.

Guerrilla marketing rule of thumb: As a guerrilla, you should be investing about 10 percent of your budget marketing to the universe — to every single person in your marketing area. Your purpose in marketing to them: to move them into your second guerrilla market.

Invest 10 percent in the universe

YOUR SECOND GUERRILLA MARKET: YOUR PROSPECTS

This market is considerably smaller than your first market, yet potentially far more profitable. The people in this market have differing degrees of interest in your offering. Some are mildly interested in products or services somewhat like yours and perhaps in yours specifically. Other members of this market are blazing hot prospects for what you are offering. They are right on the verge of buying and merely need a gentle — or not so gentle — nudge.

Guerrilla marketing rule of thumb: Nudge the people in **Invest 30 percent in your prospects** your second market by investing 30 percent of your marketing strictly in them. Interest them, inform them, intrigue them, fascinate them, woo them, do all you can — knowing that's a big and rewarding job — to move them into your third market. That's where the fun and real profits are.

YOUR THIRD GUERRILLA MARKET: YOUR CUSTOMERS

Let's take a moment to experience the warm, fulfilling sensation of satisfaction that comes while we both pay homage to the most revered of your three markets — those blessed and wonderful people who are nesting happily amidst your customer list. These are the people who make you tick. Each one represents multiple sales to you. Each represents the initial sale, when they were gloriously transformed from a prospect to a customer. That's like the transformation from infidel to true believer. Each also represents a potential repeat sale, then another, and another after that.

Each of these fine, upstanding, lovable souls also represents, if you go about things with the wiles of a guerrilla, a potential referral sale, then another, and another after that. And all those referral sales can lead to more repeat sales and more referral sales.

Invest 60 percent in your customers Guerrilla marketing rule of thumb: Invest 60 percent of your marketing funds in your own customers. Naturally, if you are opening a new business and have no customer list, you can't very well devote 60 percent of your marketing budget to your customers. In that case, devote 10 percent of your budget to your first market and 90 percent to the second. In due time, you'll have a third market.

If you already have the most valuable part of the equation, a long customer list, your marketing costs will be far lower than if you had to rent the list from someone else. Since all the people on that list already know how to buy from you, whether at a store, by mail, or by phone, it's going to be relatively easy for them to buy from you again. Since all the people on that list received satisfaction plus during their last business encounter with you, there is a great likelihood that they will buy again and recommend your offerings to others.

Never forget that the sacred obligation of every guerrilla is to move members of the universe into your prospect market, move members of your prospect market onto your customer list, then market like crazy to everyone on that customer list. **A guerrilla's sacred obligation**

If you have a product or service that does not lend itself to repeat sales, you'll have to forsake these guerrilla rules of thumb and make up a few of your own. But even you should be devoting a fair portion of your budget to existing customers simply for the referral business they can mean to you. You've heard of word-of-mouth advertising? Well, these people are the mouths. Your marketing to them will supply the right words.

Apathy toward a customer list, or worse yet, failure to maintain a current customer list, is a sickeningly commonplace omission on the part of most businesses. This means that it is also a fruitful opportunity for you. Why don't these businesses regularly make offers by phone or mail to their existing customers? Because they are not guerrillas. Few business people are. That's why your guerrilla marketing attack will bring grins to your face and greenbacks to your finances.

Guerrillas massage and fondle their customer lists and guard them fiercely. They add to them. They mail to them. They keep in touch with them. Sometimes their mailings are not oriented to making sales. Instead, they are strictly devoted to intensifying the relationship. Such mailings might include birthday cards and Christmas cards. These guerrillas know that the more intense the relationship, the more repeat and referral sales will result.

Is it possible to market to all three guerrilla markets at once?

Yes, it is not only possible but desirable to run marketing that emphasizes your product or service while enlarging the audience in your market at the same time. For example, if you own a satellite TV store, you can sell dishes hand over fist out of your own store while also extolling the benefit of satellite TV system ownership. This one marketing effort can be motivating your guerrilla market number two — those people already vaguely interested in earth station ownership, *and* guerrilla market number one — absolutely everyone in your market area.

Guerrilla fact: Customers tend to read ads run by businesses that they patronize. They do this to help rationalize their purchases from the business and their relationship with the business. So you can be sure of attracting members of that beloved guerrilla market number three. In addition, every time you put out the word about your business, increasingly large numbers of people move from market number one to market number two. And that's a much better place for them to be — the second-best place possible. This understanding of all your markets is necessary before you can mount your attack with a high and realistic expectation of permanent victory in the marketplace.

7
A Dozen Tactics to Consider

ARMED WITH YOUR WEAPONS, your insights, your plan, and your calendar, the time to launch your attack is approaching. And you're ready to move, right?

Not right. To achieve the end result of profits through guerrilla marketing, you've got to act like a guerrilla. You've got to use every possible weapon in your arsenal, and you've got to use them with tactical precision and creativity.

When you do, you have the awesome power to move large chunks of humanity from the universe to your customer list. To do this requires that you pepper your guerrilla strategy with specific guerrilla tactics — little-known, high-potency marketing efforts. This will make your attack an all-out attack — the kind guerrillas must wage, but not necessarily with large expenditures of money. Instead, the hallmarks of these tactics are patience, imagination, time, and energy. **Launch an all-out attack**

Once you know these tactics, you'll recognize that you'll be able to put some to work for your own business immediately. You can get into a launch mode by the end of this chapter. If you're going to employ the tactics, I recommend that you not wait too long or your competitor may beat you to the attack. With the influx of small businesses in all categories of products and services, only the most aggressive marketers will survive.

It's a good thing you're a guerrilla. You're aggressive. You've got the flexibility to move fast. You're not hampered by the bureaucracy that impedes by nitpicking — if you think too many cooks spoil the broth, too many marketing opinions spoil a whole lot more than a pot of soup. You're not bogged down by tradition. You're able to graciously survive with a slow but healthy and growing return. There are tactics that can help you along this path to prosperous survival while you maintain a constant attack.

How many guerrilla tactics are there? Thousands. Probably millions. As your company prospers using guerrilla tactics, you will spot new tactics, invent others, be inspired by still others. The possibilities are endless. To get your guerrilla juices flowing and to get you thinking like a guerrilla, I submit just one dozen tactics that merit consideration prior to launching your attack. These tactics are presented in no particular order, for there is no particular order to a guerrilla marketing attack. Activity occurs on all fronts. But the tactics were selected because they are relatively new, are rarely practiced by small businesses, and have been highly successful for the companies that have attempted them. They are more guerrilla-like than many other tactics because they rely on imagination more than on money. Consider which can help win the battle for customers for your business.

GUERRILLA TACTIC #1:
EIGHT STAMPS

Everyone is discovering the pleasurable returns of an effective direct-mail effort. So everyone else is receiving a mountain of direct mail. How do you get through this morass of envelopes, cards, catalogs, and packets? I just told you: With patience. With imagination. With time. With energy.

You can also penetrate the clutter of marketing with an envelope that has eight stamps affixed to it. Instead of mailing 10,000 envelopes at once, mail 500 at once. Rather than using a metered bulk rate, invest in the cost of a first-class stamp — 25¢ as I write this. A 25¢ stamp, especially a commemorative, will get a lot more attention than any metered stamp.

But eight stamps will get even more attention. Stick on one 6¢ stamp, two 4¢ stamps, one 3¢ stamp, and four 2¢ stamps — a total of eight stamps that cost you a total of 25¢ — the same as a boring meter imprint or standard stamp.

Who could fail to open a letter with eight stamps? When was the last time you received such a letter? If you found a letter with eight stamps in your mail, isn't there a good chance that you'd open it first?

The eight-stamp ploy is a perfect example of a guerrilla marketing tactic:

A perfect tactic

- It doesn't cost more money than using a first-class stamp.
- It requires extra time, energy, and imagination.
- It demands patience: You mail more slowly than usual.
- It increases your rate of response, thereby lowering your cost of marketing while making that marketing more effective.

Why don't more companies utilize this method of getting through the jungle of direct mail? Not enough are run by guerrillas.

GUERRILLA TACTIC #2:
EXPENSIVE ART DIRECTOR

If you're going to have graphic materials such as business cards, stationery, signs, brochures, ads, a yellow pages ad, and scads of other possibilities, you're going to be unconsciously judged by the professionalism of those materials. If they look schlocky, your business can be a veritable Versailles Palace, but it will be perceived by the millions who have not seen it as schlocky.

You certainly don't want to embarrass your business with the savings obtained by using the questionable graphics services of your printer, local newspaper, or yellow pages people. But you don't want to bankrupt yourself with a deservedly high-paid art director. How do you get the guerrilla professionalism you seek along with the guerrilla economy you need?

You hire a high-paid art director to design your overall format: how it looks on stationery, business cards, signs, brochures, ads, and a yellow pages ad, for example. You pay that art director the lofty-but-amortized fee billed to you, then you use the services of a low-priced free-lance art director who can follow the style set by the high-priced person while charging bargain basement hourly fees.

Pay through the nose — once

This is having your cake — the stunning visual identity that eloquently communicates quality and inspires confidence — and eating it, too — the comfortably low charges levied by your free-lance art director. That free-lancer will raise his or her prices enormously within a few years, so count your art blessings while you have them and keep an eye open at all times for potential replacements.

If you tackled this tactic properly, you will have given such good directions to your expensive art director that you won't need those expensive services again and can attend to your art needs economically and aesthetically.

GUERRILLA TACTIC #3:
MEDIA-BUYING SERVICE

If you're going to be doing any advertising in the media — and you certainly don't have to, but it can help in more ways than you can imagine, as you'll discover in the next chapter — there are several ways you can go about purchasing the time and/or space you'll require in the media. Those several ways can be lumped into two categories:

1. *Wrong ways*: by yourself, by setting up a house ad agency, or through an advertising agency

The right way to buy media

2. *Right way*: through a media-buying service

Before I explain why, you should understand that the cost of media is 100 percent to you as a business, but only 100 percent minus 15 percent to an advertising agency — which is how they derive their income. If you purchase through an advertising agency, you'll spend more than absolutely necessary and you'll probably pay for advertising services and overhead that do not benefit your business. But you can enjoy that 15

The house agency

percent discount by setting up your own house agency, which usually requires no more than saying you have one and placing your own ads.

If you buy your own media without a house agency (though I don't know why you would) you'll probably spend more money than you have to, you'll certainly take more time than necessary, and you'll make several media decisions for wrong, but well-intended, reasons. If you set up a house agency, you'll save the same 15 percent saved by ad agencies, but you'll be buried in paperwork and will lack access to all the media possibilities available to you.

Let a professional media-buying service select your media. These specialist firms learn your budget, your target market, and your competitive atmosphere, then they help you pinpoint the best media, and even negotiate the best price. Better still, they handle the complicated paperwork associated with media, and they check to be sure you got what you paid for. Both tiny and gigantic guerrillas are rapidly discovering the benefits of working with media-buying services. I certainly recommend that you try one. Many of today's advertising agencies no longer offer to purchase media for clients, but turn the whole shebang over to a media-buying service. Good thinking.

Such services in this relatively new industry are granted the same 15 percent discount as advertising agencies, but they generally return 10 percent to you, keeping 5 percent for themselves for their troubles. Some even return only 7 percent to you and keep 8 percent for themselves. But in the long run, they save you money.

Ask any guerrilla you meet: Media-buying services are worth what they charge for their consulting, negotiating, and administrative talents. Find them in the yellow pages.

Ask any guerrilla

GUERRILLA TACTIC #4:
PREPRINTS OR INSERTS

Some people call them preprints. Others call them inserts. They're the special advertising sections that you see stuffed into your newspaper or mailbox. About one third of the people who get them toss them away on sight. Another third glance at them or through them while they're going through the newspaper or mail. A final third take quite a bit of time checking out these inserts, saving them in many cases.

**Zero in on a
zip code**   A winning aspect of preprints is that they enable you to target geographic zones from the size of a zip code to a whole multistate region. For instance, if you've got a store and want to market to every single family in your zip code, preprints will do the trick for you. The families in your zip code who do not subscribe to a paper containing your preprint will receive your preprint in the mail. Total saturation.

A client of mine who has employed almost all the 100 guerrilla marketing weapons during his twenty-plus years as a tough guerrilla retailer tells me that preprints have been the single most effective weapon he's ever used. That's high praise from so aggressive and experimental a businessman.

There are five reasons to try preprints if you can:

1. They allow you to pinpoint a market, then cover it completely. Try five zip codes; try two. Zero in.
2. They give you the equivalent in space of four or more full-page newspaper ads. So you can give your prospects all the information they'd ever want. And there are no other ads in the insert competing for your reader's attention.
3. They last as long as a month, so you can use them monthly, gaining the effects of four weeks of ads.
4. You can test their effectiveness, using a page of coupons as a measurement. Preprints get great coupon response.
5. Time-limited offers work well in preprints. Offer special discounts, services, packages, or freebies.

GUERRILLA TACTIC #5: SAMPLES TO THE GRAPEVINE

**What
businesses do
your customers
patronize?**   All you've got to do is ask the right question — "What businesses do my customers patronize?" — and then provide samples of your product or service to the answer to this question. As with all guerrilla marketing, indeed, as with life itself, your quality must be unimpeachable for this tactic to work. But if you offer quality, you're going to love this inexpensive way of putting heaps of satisfied customers onto your customer list.

First of all, I understand that you do not, in all likelihood, own a restaurant. If you do, you are one lucky guerrilla, considering the upcoming example. But even if you are not a restaurateur, you can use the principles of this tactic to attract customers to your own business.

A restaurant opened in a metropolitan area, then distributed coupons good for two free dinners to all the hairstyling salons within a one-mile radius of the restaurant. Naturally, the stylists checked out the restaurant. Because it was as wonderful as it implied with so confident a stroke as free meals to strangers, the hairstylists loved it and talked it up in their salons. As you may know, gobs of information can be disseminated in a local styling salon or barbershop. If a restaurant is horrible, a lot of people find out about it in a hurry. If it is fabulous, the same phenomenon occurs.

This particular restaurant exhibited the pure guerrilla spirit with this maneuver: a tiny investment, a huge imagination, a happy payoff. Just what guerrillas are supposed to do. You can do it too, if you know where information transmittal takes place in your industry, where your customers pick up data, where word of mouth is most likely to occur. If you can, offer some proof of your excellence to the people who say the words — and you're on the grapevine.

GUERRILLA TACTIC #6: YOUR CLUB

Again, allow me to give you a real-life case history of a guerrilla in an industry that is probably different from yours, but certainly no less competitive. This man used his guerrilla tactic to capture an incredible 80 percent of his market within three years. Regardless of your business, you can use the concept behind what he did. A practicing guerrilla marketer who owns an immensely successful chain of video rental stores in Omaha, Nebraska, he merely invited prospects to join his video club. The concept behind it is that there is absolutely no cost to join.

What do people receive with their membership? They get

a beautiful gold-trimmed membership certificate plus six free video rentals. "After that," says the owner of the chain, "they're mine." It's no surprise that this particular guerrilla has earned the "Video Retailer of the Year" award from *People* magazine. And it's no surprise that his "club members" respond to his large assortment of marketing vehicles: signs, direct mailings, ads, TV commercials, and inserts. Sounds simple on the surface, but ask yourself how often you're given free things that you would gladly pay for. Hardly ever; perhaps never.

Who can blame prospects for joining a club that gives them items of value for no charge? Of course they feel obligated. But they also feel a sense of kinship with the store. There's a point of interest about this particular guerrilla, a belief which explains how he came up with the club idea in the first place. His motto is: "If you go into business to make money, you'll go out of business. If you go into business to serve customers, you'll make money."

One guerrilla's motto

Apply that motto to your business. See if you can come up with the ingredients for a club or something like a club, a method by which your prospects and customers can identify with you. Present them with products or services that have no strings attached. Then reap the benefits that accrue from this very guerrilla and very human tactic. If you put this tactic into the category of tithing, you've got the right idea.

GUERRILLA TACTIC #7:
MARKETING TRENDS

This is an ever-changing tactic, never the same twice, always effective, unknown to most of your competitors. It entails keeping abreast of today's marketing trends by subscribing to one or two of the marketing trade journals. There are several to choose from. Check the bibliography and select one for yourself. Better still, go to the library and look through them in person. You'll find at least one that's easy to read and that comes to you weekly.

If you go about this tactic in the right way, you'll gain at

least a one-year head start over your competitors. Here's the **Gain a head**
right way: Spend a *maximum* of fifteen minutes each week **start**
looking through the marketing or advertising magazine to
which you subscribed. Skip right over the industry gossip, the
news about companies, the hirings and firings and promotions,
and the stories about meetings and conferences. Concentrate
on the sections dealing with new trends in marketing. Out of
every four issues, perhaps three will report on trends and marketing techniques that are inappropriate for your business. But
approximately one of the four issues will describe a trend or
technique that's *exactly right for your business.*

The issue will describe the technique, the results, all the
details you'll need to dive right into it for your own company.
Don't wait! Great marketing ideas tend to be copied until virtually everyone is on the bandwagon. That's when you should
leap off.

Your job as a guerrilla is to spot and capitalize on new marketing trends and techniques before there ever is a bandwagon.
Be warned: Much of the material in the marketing publications
is downright boring. Much is just not affordable or sensible
for your company. But that one pure gold idea a month —
and that's twelve pure gold ideas a year! — that's what is going
to make you love this tactic and incorporate it as part of your
business routine, a highly enjoyable part, I'll bet. As you can
see, being a guerrilla requires keeping abreast of current knowledge — in marketing, in psychology, in your own industry.

GUERRILLA TACTIC #8: TESTING COPY

The power of headlines has never been in dispute. Albert Las-
ker, an early advertising titan who founded his own advertising
agency and wrote prolifically about advertising, said, "The
headline in the end, today as 25 years ago, is 90 percent of all
there is to an ad. Why do I say 90 percent? Because if you
don't stop them with the headline, they won't read the rest."

Knowing this, it won't be tough to convince you that you
can save a bundle if you test headlines. Test them by using **Test headlines**

local newspapers and offering gift items to people who respond by phone, mail, or in person. Test headlines by direct mail, too, mailing to just a few hundred names to get a line on your best headline.

Soon, you'll get a fix on which headline is pulling best for you. Test that finding by trying the strong headline in the weak newspaper or radio station or mailing list. If the line still pulls in the business, grin widely and take a deep breath of gratitude. The last thing you want to do is invest money in a bad headline. The first thing you want to do is invest the maximum in a proven good headline.

It's case history time. A businessman tested a headline that offered a two-for-one promotion: "Get two for the price of one." That headline brought in 850 responses. He measured that response against a headline for a one-cent promotion: "Pay for one. Get the second for one penny." That headline drew 1300 responses.

If he hadn't tested, he wouldn't have known. If you don't test, you won't know. Guerrillas know because they test. That's what gives them the faith to make their deep commitment to their marketing campaign.

GUERRILLA TACTIC #9: THUMBTACKS AND STAPLES

In virtually every populated area of the United States, in fact in most civilized areas of the world, are bulletin boards that serve as a forum for signs advertising entrepreneurs, businesses, products, and services of all kinds.

Bulletin board power   Where I live, the San Francisco Bay Area, there are over eight hundred public bulletin boards. Many small businesses post their signs and circulars on these bulletin boards on a regular basis. They can't get to all eight hundred, but they can get to many key locations. And the business they derive from their postings is considerable, especially when you realize how little money they invested in marketing. Like all guerrillas, they invested patience, time, energy, and imagination. And like all guerrillas, they are receiving a healthy return for their efforts.

Bulletin boards are found in laundromats, supermarkets, bookstores, campus centers, school hallways, employee lounges, waiting areas, and myriad other locations where people happen to congregate.

This marketing medium is so hot these days that there's a Bay Area company that exists solely to service people who want their signs on the bulletin boards. The company is called the Thumbtack Bugle. Corny? Well, their main competitor is called the Daily Staple. These companies, for far less money than you'd think, not only post your notice or circular on any or all of the eight hundred bulletin boards around, but they also distribute your fliers in stores and shops. They post them in windows, hand them out at concerts and events, even during the rush hour. Once they know who your target audience is, they know how to reach them with your circular.

These days, the Thumbtack Bugle, easily reached by calling (415) 221-6666, also helps you plan your message, mails brochures and postcards for you, designs those very same brochures and postcards, and even gets your message into the media as free publicity: They write press releases for you and mail to over 150 media desks.

Even if there is no similar service in your area, you can put the idea to work for you. The cost is low. The investment is only time and energy. The results are proven effective for all sorts of businesses. Might yours be one of them? A true guerrilla finds a way to answer that in the affirmative.

GUERRILLA TACTIC #10: SOLVING PROBLEMS

Life just seems to be easier for businesses that position themselves as problem solvers. It is an axiom of American marketing that prevention is the hardest to sell and solutions are the easiest. So try to present your product or service as the solution to a problem. It is not difficult to flag the attention of prospects with the problem; merely state it clearly. Then, offer the relief

Customers love solutions

and joy of your product or service to those with the problem.

Former advertising superstar Alvin Eicoff said, "Set forth the problem. Explain the solution. And then demonstrate why your specific product or service best provides that solution." He also said, "The first visual and audio elements of a commercial should state the problem clearly and concisely. The potential customer should feel a strong personal identification with the problem presented, reflexively nodding his or her head in acknowledgment."

Involve your prospects

You can apply this advice to many other forms of marketing such as your brochures, telemarketing scripts, direct mailings, signs, and sales presentations. Your prospects will be more likely to buy if they are involved with your offering. You can be sure that if your offering solves their problem, they will be quick and easy to involve. You can also be sure that involvement creates momentum — that important momentum that fuels the marketing process.

Problems abound: too fat, too thin, too small, too large, too poor, too bored, too slow, too busy, too stressed out, the list goes depressingly on and on. Surely there is at least one problem that your business can be positioned to solve with certainty. A guerrilla could find one. I bet you can, too.

GUERRILLA TACTIC #11: BRIBES WORK

If you want to enjoy a dramatically increased response to any of your offers, try offering a free gift to anyone who responds. The free gift can't be some junky item that looks and feels cheap. Instead, it should have a high perceived value. These days, with the cost of electronic gizmos such as calculators, wristwatches, desk clocks, clock pens, and more coming in at around $2.50 each, and less in large quantities, you should give serious consideration to bribing your prospects with freebies. Tell them that if they send for your brochure or call for more details or set up an appointment or come to your place of business, they will be given a free gift.

Everybody loves free gifts. I was once engaged in a project

for a huge bank that was attempting to solicit the accounts of people with net worths of over $1,000,000. A mailing offering a free brochure elicited a response exactly one-sixth the size of a mailing offering the same free brochure plus a leatherette memo pad (value $1.19).

The most successful direct-mail user I've encountered claims that his company's responses have increased over 100 percent since they've been offering bribes to anyone who requests a brochure. Even better, that response rate goes up even more if they show a *full-color picture of the bribe* on the mailing envelope. They obtain their very best response by showing that same picture on a postcard — proving the efficacy of both bribes and postcard mailings. Postcard mailings will be discussed in detail in Chapter 9.

You can get a line on the kind of bribes I mean by requesting a free brochure from Schmidt-Cannon at (818) 961-9871. Before leaving this tactic behind, I urge you to offer bribes to existing customers who refer new customers to you. Talk about a heartwarming response!

Something free for you

GUERRILLA TACTIC #12: FIND MULTIPLE USES

If you're going to invest in a fine photograph or illustration for a marketing weapon such as an ad or brochure, spend the money necessary to gain the quality you want, then amortize that cost by using the photo or illustration in as many places as possible.

Down goes the cost

You may need the photo or illustration for an advertisement. But no true-blue guerrilla would stop there. The guerrilla would use the artwork in a brochure, possibly for the cover. He or she would blow up the art — I mean really blow it up big — to use as a poster. It would also be valuable as a sign, either for wall or window or both — even for a billboard.

Some of my clients have blown up their artwork to a full 4-by-5-foot A-frame sign, utilizing it outside their place of business as well as inside. Others have used their art on their business cards, on catalog sheets, as minor parts of future ads, as the basis for calendars, in trade show displays, in preprints, and as part of a press kit. As you can see, the opportunities are bountiful, and the more you use, the lower the cost for the photo or illustration. With each year you use the art, the cost goes down even more. Soon, the expensive photo or illustration becomes one of the most satisfying marketing values you've ever invested in. It works that way with guerrillas. It should work that way with you. Don't save money foolishly by using a cheap photographer or illustrator. Use the best. Then find multiple uses for the fruits of their talent.

This is a guerrilla tactic through and through. The financial investment is not as large as the investment in imagination. With enough imagination, the financial investment appears tiny and insignificant, while the results of the investment appear rich and rewarding.

Many of these tactics are ideal for today's marketplace, but may be outmoded or improved on within a few years. So this is no time to dally. Right now — before you launch your attack — determine which of these tactics you can use. As a right-thinking guerrilla, you should consider them all, then use as many as possible to increase your profits while coming up with another dozen tactics of your very own. After all, that's what guerrillas are for.

8
Gaining Effective Media Support

ALTHOUGH MANY GUERRILLAS will employ weaponry that does not require the power of the media, the majority of business people will want to enlist the aid of the media, which influence countless people, to add firepower to their marketing. You can use one or more media to extol the benefits of your offering to hundreds, thousands, even millions of your prospects. However, since not all the media reach large groups of readers or viewers, many guerrillas will use media for several other reasons:

1. *Credibility*: Many readers of certain publications trust the advertisers in the publication as much as they trust the publication itself.
2. *Targeting*: If your offering is not for the general population, there's a good chance that a specific medium gives you a more accurate aim at your target market.
3. *Politics*: Frequently, it makes sense to use the media because they help your company become accepted by your own industry as well as by the public at large.

Perhaps the biggest mistake made concerning the media is to overlook them entirely, figuring that their cost doesn't fit into your plans. Many a guerrilla has taken advantage of this naive attitude by making a big splash in a major medium, gaining enormous profits because competitors were intimidated by the media, indeed, by the whole idea of marketing.

The biggest mistake

For instance, with the growth of cable TV and, with it, lower-than-ever TV rates, the idea of television advertising should no longer repel small advertisers. Instead, it should attract them, because of the low rates and the universal acceptance of the truism that TV generally rules the roost in the mass media.

Another grave error is to fall prey to a smooth-talking media salesperson who knows how to sell anything to anyone, including selling an inappropriate marketing medium to you. If you're selling tickets to a rock concert, a classical music station is hardly the place to advertise them. If you're selling computer equipment, a senior-citizen newspaper is most likely going to miss your target audience.

About the only thing worse than not using the media at all is to use the wrong ones. I once had a client who was enamored of television. She loved it so much that she advertised on it even though she could afford only one commercial per week. Well, one commercial per week is not going to sell anything. You may as well contribute that sum to your favorite charity rather than underbuying a medium that demands far more frequent exposures than one commercial.

Of course, if that one commercial is going to be seen on a show that will be watched by two hundred million people, that's a different story. But that's also a different cost, too. Rather than talking $25 for one minute, we may be talking $250,000 for thirty seconds. To many advertisers, that may be a terrific bargain — only $8 per thousand viewers. Still, one advertising impression, thirty seconds long, is not going to woo many of the two hundred million people into buying what you're selling.

Expensive and inexpensive media   As students of guerrilla marketing know very well, whether or not a medium is expensive or inexpensive has little to do with its price. If you spend $10 on an ad and nobody responds to it, you have paid for a very expensive medium. If you spend $10,000 on an ad and you earn $50,000 in profits, you have paid for a very inexpensive medium. As you can clearly see, the price does not reflect how expensive or inexpensive the medium was.

To a guerrilla considering the media, the value of a medium is measured by its potential effectiveness, not by its dollar amount. Guerrillas know there are only two kinds of marketing:

expensive and inexpensive. They know that expensive marketing is the kind that doesn't work, regardless of its price. So don't be bamboozled by a low dollar amount. Ask about results the medium has achieved for others in businesses similar to yours. If you can't get a straight answer from the sales rep, make a few calls to advertisers. They'll level with you — unless they consider your business competitive with theirs.

Still another major media error is using too many of them. **Don't spread** Using many media is wonderful if you're McDonald's. But **yourself too** since you probably aren't — yet — keep your list of media **thin** short, limiting it only to those media that you can use properly and which have demonstrated their effectiveness to advertisers such as you.

You can easily tell how effective a medium is by the length of time an advertiser has been running ads there. If you've seen a company such as yours running an ad each week for three years, you can be confident that the medium is a winner. But it will only win for you if you use it long enough, if your ads are interesting and motivating enough, if your ads inspire a feeling of confidence in you by the readers or viewers of the medium, if you make the right offer, and if you say the right things. It sounds like a tall order and it is a tall order, but if marketing were simple, there would be no sidewalks, since everyone would be driving their Rolls-Royces rather than walking.

Because I don't want you to expect attack support from the media, only to be frustrated because it doesn't materialize, I **Classic media** am focusing at first on the classic media errors made by busi- **errors** nesses of all sizes.

High on the list is not using a medium long enough. You may have discovered the perfect medium. It is priced right; it reaches the exact people you want to sell; it has worked for advertisers such as you. But you run your ad in it and nothing happens. The right thing to do is to continue running your ad there. Remember, even brilliant marketing rarely works instantly. The wrong thing to do is to abandon the medium. As members of the human race, we are accustomed to instant gratification. We want to run our ad on a Monday, then make a huge bank deposit on a Wednesday, thanks to all those sales we made on Tuesday.

Ranking right up there on misuses of the media is selecting the optimum medium, then blowing more money than you should on a huge ad — thereby preventing you from keeping your name and offer in the public eye on a regular basis. If you can afford to run a huge ad, by all means run it — but **Consistency** not if it means you'll become invisible for the next few months. **over impact** Consistency will prove a more valuable ally than impact. The combination of the two is the ideal. But if you have a choice, choose consistency over impact every time. That choice becomes quite simple if you realize that the choice is really between impact and profitability.

To determine how big your ads should be is a simple matter. If you're to use the media effectively, the ads should be big enough to tell your story and/or show your feature. Invest in the space necessary to accomplish those aims. Ads that are too small might deprive interested prospects of salient facts and fail to give them enough information on which to base an intelligent decision. Too large an ad might be a waste of your precious marketing budget.

How do I figure how much I should invest in the media?

There are six options available to you. First of all, be sure you have enough of your budget allocated to nonmedia marketing before appearing in the media. That means having first-rate stationery, business cards, signs, telephone answering, brochures, sales training, market research, and all the other elements that guerrillas find so crucial to their arsenal. These are thoroughly investigated in *Guerrilla Marketing*. The selection of weapons is large; many of the weapons are free. Just be sure you attend to these non-media details before you allocate funds for media. The last thing you want, for example, is to run ads that result in a blizzard of phone calls and for the calls to be answered by a tactless, uninformed grouch.

Then, choose one of the following six methods of determining how much to spend. The first five have inherent problems, as you'll see. So I've saved the best for the last, listing the others in descending order of their popularity among small businesses in America.

1. *Spend whatever you can afford:* This ever-popular method tempts small and new business owners because it appears so sensible, affordable, and simple. But in terms of marketing and media, it is nonsense. Using media in a wimpy fashion keeps people away from you, some who patronize regular advertisers while you're away. It encourages impulse buying of media *after* good months, and discourages any media buying *after* poor months. It gets in the way of consistency, a guerrilla marketing calendar, and consumer confidence. Guerrillas shun it, but secretly pray that their competitors practice it.

2. *Keep up with the competition:* This method forces you to react, not lead. The opposite is a better approach: Stay ahead of the competition. And as long as you're there, don't be predictable. Build this unpredictability into your marketing calendar so that your competition can't keep up with you even if they want to. Guerrillas aren't satisfied with merely keeping up.

3. *Allocate a percentage of your past sales:* This is simply determining a percentage to invest in marketing and basing it on the past, rather than the present and future. The big problem with this commonly misused method of determining a media budget is that it impedes growth. Sure, it's simple to pick a percentage and stick with it, but if you look backward to determine that dollar amount, you are looking in the wrong direction. Guerrillas look the other way.

4. *Use the cost-and-objective approach:* Doing this requires that you determine your media needs and objectives, calculate the cost to attain them, then spend your dollars where they are needed and can positively affect sales the most. But if your plan is too ambitious for the funds you have available, you can be in serious cash trouble. If you scale down your plan, it may

force you to use a selected medium improperly. Guerrillas like spending money where it's going to work, but hesitate to take the risk of overspending.

5. *Use the per-unit approach:* Here, you treat each of your products or services as a separate business with a separate budget. From each budget, you allocate a specified amount for media. This gives you an accurate picture of your media effectiveness, but can be very complex and time consuming. Guerrillas employ this method only if they have a basic offering without too many stock-keeping units.

A guerrilla's favorite 6. *Invest a percentage based on your projected sales:* This is my personal recommendation, based on real-life experience with large, small, new, and old companies. It forces you to look into the future, to act instead of react, to make and keep a commitment, to be consistent, to force yourself to earn confidence from your prospects. What size percentage should you invest? In 1987, the average U.S. business invested about 3 percent. Marketing experts agree that the minimum should be 5 percent. Guerrillas have proven that the minimum should be 10 to 15 percent.

Brand-new businesses may have difficulty projecting sales with accuracy, so for their first six months, the cost-and-objective method should be employed. Decide what media you'll need for six months, and that cost is your media budget. After six months, you can become more scientific.

Invest more at first As a rule of thumb — even a rule of heart and brain — new businesses must spend more at first than later on, when they're established businesses. But take heart: All existing businesses were once new and had to spend a lot. Feel even more glee if they *didn't* spend a lot. That may leave the door open for you to experience instant market domination, a heady feeling for any business person, guerrilla or not.

How do I know which media to select
for my attack?

Every day, there are more new media coming down the pike: new cable outlets, new radio stations, new newspapers, new magazines, new forms of media, such as signs on parking meters. Which to choose? Not easy to answer. But there are guidelines:

See where your competitors are advertising. That usually gives you a good fix on where you might do it, too. Prospects already know that advertising for your sort of offering appears in this type of medium, so the stage is set. Also, you can meet your competitors on a more face-to-face basis this way.

Talk to others in your industry; their ideas can help you. At trade shows, conventions, and in industry publications, you can generally get a good idea of what media are working for offerings such as yours. At just a single trade association meeting, you can learn some special media secrets that can nourish your bottom line for years to come.

Test until you are convinced you've selected the right media. Of course, this will involve testing some wrong media along the way, but it's a wise investment because the information can be worth a fortune to you. Sometimes, you can test two headlines at the cost of one ad. This luxury is yours if the medium offers what is termed an *A/B split.* That means, say, in the case of magazines, that half the issues will carry one of your headlines and half will carry a different headline. Not all publications offer A/B splits, but if those that interest you offer them, take advantage of this low-cost method of real-life testing. You gain hints about the medium and certainties about your headlines for the same money.

Don't overlook any of the media; the more you test that you can use properly if they work, the better for you. Maybe your competitors haven't experimented with telemarketing, with postcard deck mailings, with satellite TV advertising. Sure, advertise where your competition advertises. But be sure also to venture where no competitor has ventured before — such guerrilla treks make sense when practiced with restraint.

Think in terms of narrowcasting, not broadcasting. You don't want to waste your valuable media budget talking to people who will never be hot prospects for what you sell. So narrow down your list of media to reach your prospects only, then talk only to them. You shouldn't care how many people are exposed to your message, but you should care like the dickens how many prospects are exposed to it.

Don't worry about the CPM; keep your eye on the CPP. Some marketing people revere the phrase *CPM*, which stands for "cost per thousand" — the total cost to tell your marketing message to one thousand people. Guerrillas care not one whit about CPM. Instead, they are interested only in the *CPP*, which stands for "cost per prospect," and they are far more interested in prospects than people in general. This interest helps them save money, not waste it.

Keep an eye open for fabulous bargains in media. These are more available than you might imagine. Some media might be willing to engage in a barter arrangement with you, even a three-way barter with another party. Others may have a blank page left over where they didn't sell ad space, so they might be willing to let you have it for a song — plus a check, but not as large as they normally charge for a full-page ad. Still others may by open to a "per inquiry" (P.I.) or "per order" (P.O.) arrangement. This enables you to get the advertising space at no up-front cost. However, you compensate the publication with a preagreed amount for each inquiry or order the ad elicits. Hundreds of publications and electronic media are open to this, but few have clear policies, so they make their decision to go along on a one-by-one basis. If they feel they can earn more by getting a piece of your action, they'll give you the space or time. They make this decision based on your product or service, your price, your ad or commercial, and their past experience with P.I. and P.O. arrangements.

The changing media In *Guerrilla Marketing,* I examined all of the media available to guerrillas. But the media picture is constantly changing, primarily in ways to help guerrillas, so it's a good idea to look at the current media possibilities for your company. Here are the new insights and advances that have occurred since you last investigated the media.

Realize that every medium has strengths and advantages unique to it. Each has disadvantages, too. As you explore the media options available for you, always be asking: "Can I use this medium properly?" The idea, as any guerrilla knows, *is to use as many of the media as you can use properly and as will be appropriate to your business.*

Since there are so many media these days, I have divided them into three categories — mini-media, maxi-media, and non-media — just as in *Guerrilla Marketing*. A victory-minded guerrilla will employ weapons from all three categories. Let's see how many are right for you.

The Mini-Media

One associates the small business with the mini-media. These include canvassing, personal letters, telephone calls, circulars, brochures, classified ads, the yellow pages, and signs. Some of these marketing vehicles are usually not even considered by many large companies. It's hard to imagine the president of a multibillion dollar conglomerate running a classified ad. Yet, some members of the *Fortune* 500, guerrillas in three-piece suits, are discovering the awesome power of a single personal letter written to a key customer. Many know the merits of a phone call. Guerrillas of all sizes can augment their attack with the mini-media. Don't be misled by the prefix "mini." The only thing truly mini about the mini-media is their cost. The profits they create can be and have been impressively maxi.

Maxi-profits from mini-media

CANVASSING

Canvassing can be accomplished door to door, store to store, trade show booth to trade show booth, motorist to motorist, boatowner to boatowner — in limitless ways to a fertile imagination. The greatest strength of canvassing is eye-to-eye contact. The first few seconds of a canvass set the stage and should be accompanied by eye contact, a smile, and, if at all possible, the name of the prospect. The best contacts have nothing to do with business, buying, or selling. They establish the human bond. Canvassing becomes a lot more successful if visual sales aids are used. Selling points made to the eye are 68 percent more effective than the same selling points made to the ear. The attire of the person canvassing is almost as important as the product being sold. The prospect must buy the canvasser before the prospect buys the offering. The cost of canvassing is the cost of your time if you do it yourself or the cost of the commission if you hire others to canvass. It's an economical weapon in the guerrilla marketing attack.

PERSONAL LETTERS

Personal letters are not personalized letters, but letters from one person to one other person. The greatest strength of personal letters is their ability to make many personal references. You can talk about the person and the person's life — probably that person's favorite topic. Be sure your personal letters look, act, and feel like personal letters. Details and examples can be found in *Guerrilla Marketing*.

TELEPHONE CALLS

Don't be a stranger

These are personal letters over the telephone in that they are warm, human, and loaded with personal references. The greatest strength of telephone calls is their ability to help you establish rapport. Telephone calls are far more effective if the person you call has received a letter from you or heard of you through the media or some other way. A prime advantage of telephone calls is their ability to enable sellers to establish contact, make a presentation, and close a sale. This gives local companies national potential. That's why the telephone can be a potent weapon in your guerrilla arsenal.

CIRCULARS

Circulars are different from brochures in that they give less information and are less ornate. The greatest strength of circulars is their economy. They can be distributed on street corners, under windshields, at events, on bulletin boards, and more. They're most effective when making an announcement, granting a special price, presenting a time-limited offer, or tying in with an existing promotion.

BROCHURES

Brochures provide a great many details of your offering and should contain an order form and phone number to call and order through if feasible. The greatest strength of brochures is their ability to convey enough information for a prospect to buy. Brochures enable guerrillas to run small ads in many publications, each ad highlighting the key benefits of their product or service, then offering a *free* brochure. This process is referred to as the "two-step" in *Guerrilla Marketing*. Each person who requests a brochure goes onto a guerrilla mailing list. Between 25 and 33 percent of people requesting a brochure will eventually make a purchase from the brochure — provided the brochure isn't a disaster. Brochures should not be freely given away in a retail environment because they give a prospect an excuse not to buy right then and there — "I'll just take this brochure home and look it over."

Brochure requesters become buyers

CLASSIFIED ADS

Classified ads reap dynamic results for guerrillas running them in newspapers, magazines, and newsletters. A classified-only TV station recently made its debut in California. The greatest strength of classified ads is their ability to reach honest prospects. Most people reading any particular portion of the classified section are seriously interested in buying. Classified ads reach prospects with such accuracy that they often outpull display ads costing far more. This is especially true since the advent of new classified sections such as Antiques, Travel, Computer Equipment, and several more. Classified ads can be part of an expensive but highly profitable mail order campaign if you take advantage of the high circulation of some of the national newspapers and magazines with classified sections.

New classified sections

THE YELLOW PAGES

"Yellow pages" now refers to the many directories that compete for the consumer's loyalty and attention in addition to the yellow pages directories. The greatest strength of the yellow pages is their ability to let you meet your competition head-on. A small, new business can display the pizzazz of a large, established business, using exactly the same size ad. Don't spend money in the yellow pages if your competitors aren't there. That means the public doesn't look for offerings such as yours in the yellow pages. The color red, while adding appreciably to the cost of a yellow pages ad, pays for itself and then some, more often than not, according to most reports I've heard. If you obtain a large amount of business through your yellow pages ad, you might want to invest in more ads in other sections of the directory, other areas in your locale, and other directories. Give as much information as possible, within the principles of attractive design, in your yellow pages ad. I promise you that readers are hoping you'll give information. Your ad should reflect your overall marketing strategy. In addition, show the credit cards you honor, give directions (and a map, if necessary) to your place, and mention brand names to enhance your credibility. Guerrillas rise to the fore in the yellow pages, where the battle wages furiously on each page.

When not to be in the yellow pages

SIGNS

Mini-media signs appear on walls, in windows, in store aisles, on bulletin boards, on A-frames, on shopping carts, inside and outside, and in a wide range of sizes. The greatest strength of signs is their ability to capitalize on impulse reactions. Earlier I mentioned the American need for instant gratification. In 1986, a full 64.8 percent of buying decisions were made in the place of purchase. Signs can be used to direct prospects to your location, to give information, to announce special prices, or to aid salespeople in their sales presentations. Guerrillas running busy operations create signs with enough information to answer most questions posed by customers. Referring to these signs as "silent salesmen," they frequently frame the signs handsomely, treating them as store decor although they are unquestionably marketing vehicles. Signs are so effective that a new venture allowing national advertisers to place signs on supermarket carts netted the company $50 million in sales after just two years in operation. No wonder. Signs work.

Generate impulse reactions

What signs can do

The Maxi-Media

If you read *Guerrilla Marketing*, you'll know that by the maxi-media, I mean newspapers, magazines, radio, television, billboards, telemarketing, and direct mail. Don't be intimidated by the maxi-media. Recognize that they've been changing in ways to help your company. All are more financially accessible, yet as powerful as ever — with the increased potency that comes from using more science and less prayer these days. You can gain important data on all the maxi-media in *Guerrilla Marketing*. The following information is relatively new.

Don't be intimidated

NEWSPAPERS

Newspapers range from national to metropolitan to neighborhood, from campus to ethnic to trade, from daily to weekly to monthly, from classified-only to shopper to business. All should be considered. The greatest strength of newspapers is news — so present your information in a newsy manner to capitalize on the mind-set of the reader. Guerrillas in droves are being helped by the inception of newspaper zone editions. These are regular newspapers with special sections targeted to small geographic areas in the market. For instance, as I write these words, the cost for a column inch of advertising space in the *San Francisco Chronicle* is $97.17. An advertiser would have to spend $971.70 to run an ad two columns wide and five inches high (ten inches total). Worse yet, the vast majority of people reading the ad would be too far from the advertiser to patronize the business. In large cities, this has tended to eliminate newspapers as a viable marketing medium for local businesses.

To the rescue: zone editions. The *San Francisco Chronicle* now publishes five zone editions. The cost for one column inch of advertising space in one of the zones is only $10.21. Now an advertiser has to spend only $102.10 for that two-column-by-five-inch ad. And almost everyone reading it lives or works close to the business. The ads run in special sections composed of standard editorial matter and local newspaper ads. Check to see if zone editions exist in your area. If they do, take it from me — they help move merchandise and generate profits for guerrillas.

Guerrillas report that the use of a color is a wise investment and more than pays for itself. But remember that you might have to run a large ad in order to qualify for color, and the cost of the ad and the color might be too high for you to invest on a weekly basis. Spend an hour or so with a salesperson from the newspapers in which you are interested. Ask them to help you succeed with newspaper advertising — and stay within your budget.

MAGAZINES

There are two categories of magazines for you to consider: consumer magazines and trade magazines. Of the consumer magazines, some are national and others local. The greatest strength of all magazines is reader involvement. Another major advantage offered by magazines, perhaps more than by any other medium, is credibility. The credibility the readers attach to the magazines is unconsciously bestowed on you. Anyone who thinks they can't buy credibility might consider ads in prestigious magazines aimed at their prospects. **Credibility for sale**

Still another reason to consider magazines as a potential weapon is their powerful reprint value. You can run a full-page ad in 1989 and continue to mail reprints in 2089. Each reprint would say, "As advertised in *Time* magazine" — or whatever publication you selected. **A 100-year attack**

These days, magazines allow you to aim at clearer targets. Instead of settling for the audience who reads newspapers — broad and vague — you can direct your message to prospects for your offering: photographers, gardeners, cooks, skiers, hundreds of special-interest groups. You can dramatically enlarge the number of magazine readers requesting brochures from you if you run your ads in those magazines that offer "bingo cards" at the back. Those are the cardboard pages with a host of numbers. Readers are asked to supply their name and address and to circle the number of each advertiser from whom they wish more information. I've seen a quarter-page ad in an airline magazine elicit 1247 requests for brochures. You can gain the credibility of magazines at a fraction of what you think they cost if you run your ad in a regional rather than the national edition of a magazine.

We have Seth Godin and Chip Conley, authors of *Business Rules of Thumb* (Warner Books, New York, 1987), to thank for publishing new findings about magazine advertising. They turned to the famed research firm founded by Daniel Starch to give us ten rules that have been discovered through scientific research. **Ten rules**

1. A two-page spread attracts about 25 percent more readership than a one-page ad.
2. A half-page ad is about two thirds as effective as a full-page ad.

3. A full-page color ad attracts about 40 percent more readers than a black and white ad.
4. Multipage ads attract more readers than single pages or spreads, but not in direct proportion to the number of pages involved.
5. Position in the front or back of the magazine (except for the covers) does not matter.
6. Readership does not drop off when an ad is rerun several times in a magazine.
7. Photographs are more effective than drawings. I wish I had known that during my mega-agency days when my fellow employees and clients needed that information.
8. Illustrations showing the product in use are better than static product illustrations.
9. Ads with people in them score higher in readership studies.
10. Black and white ads are about 20 percent more effective than ads with black and one color.

RADIO

Foreground versus background

For purposes of a solid guerrilla marketing attack, think of radio in two formats: foreground and background. Foreground radio includes news shows, talk shows, public radio, sportscasts, and religious shows. Because it's verbal, foreground radio requires active listening, and commercials are often integrated into the shows. Background radio includes music shows, which require less active listening. Commercials are more of an intrusion than on foreground radio — unless they are matched to the musical tastes of the listeners. In research questionnaires, try to find out what kind of radio stations your customers listen to.

The prime strength of radio marketing is its intimacy. Usually, the radio is in a one-on-one relationship with the listener, who is generally driving, or perhaps working at home. This intimacy promotes a more personal relationship than is offered by printed pages. Many advertisers report happy results from remote broadcasts held at their place of business. An air of excitement is generated by the on-air personality, and people come to get in on all the hoopla, and hopefully to buy whatever you are selling.

Another advantage of radio is that you can zero in on senior citizens, women, kids, men, cooks, office workers, and devotees of rock and roll, country, oldies, jazz, heavy metal, reggae, folk music, middle-of-the-road, or easy listening. Still another advantage is that you'll find about five radios in the average American household, plus another in each car. A most ubiquitous medium. And very flexible.

Should you have music in your commercials? If you're going to give them a lot of air play you should. One highly successful radio expert claims that music makes radio 33 percent more effective. But on a live radio talk show, the music would be an unconscious introduction to the listener. It would say, in its own language: "And now a word from our sponsor." I'm not too happy telling you this, but studies from Columbia University indicate that the faster the announcer speaks, the more attention the listeners will pay and the deeper will be their degree of comprehension.

Fast talking is good talking

TELEVISION

There is no question that television is the undisputed heavyweight champion of marketing, although some direct-mail marketers might look askance at the word "undisputed." What is never disputed is TV's ability to talk to eye and ear, using words and/or music, to reach both left- and right-brained people, and to capture attention in a jiffy. I'm not saying that TV does do that; I'm only saying it can. The paramount strength of television is its ability to demonstrate the efficacy of your product or service. The print media can show your offering in action, but not with the verve and excitement of television. TV can show the before, the after, and the during.

These days, television advertising includes, in order of costliness, network television, spot (local, single-market) television, cable television, and satellite television. In some instances, cable TV is the lowest cost TV anywhere: $3 for a thirty-second commercial on certain cable systems. So if you thought TV was something reserved for big wheels, realize that you can probably afford to start using it to become one of the big wheels yourself. In fact, the biggest change that has taken place in television since *Guerrilla Marketing* was written is the plummeting cost of TV commercial time. TV costs can be further reduced by five factors:

1. Package prices
2. Agency/Service commissions
3. Multimonth contracts
4. Tough negotiating
5. Desperation to sell as the time draws closer

How to think of TV

To use television and gain the greatest benefits from it, think of it as a visual medium with audio enhancement. You must tell your story with visual means, using the audio to aid and abet your tale. You're a goner if you think of television as radio with pictures. These days, it makes more sense than ever to show the name of your business at the beginning, in the middle, and at the end of your commercial. The reasons: (a) About 65 percent of people own videocassette recorders, and 80 percent of them use their remote controls to fast-forward through the commercials. Viewers see but don't hear them when this happens. (b) The vast majority of TV viewers are now armed with a remote control with which they mute the TV set when the commercials roll. That's why it makes guerrilla sense always to keep your name on screen.

Although in 1986 the average thirty-second TV commercial cost $93,000, you can produce a powerful commercial for less than $500 if you have a good offer, a clear benefit for the viewer, a script, rehearsals for all parties who will be involved in the spot, a tight preproduction meeting, and a commercial that requires no voice on camera. By shooting silent footage and combining it with a prerecorded soundtrack, or by recording the sound later, you save scads of dollars. Folks, we are talking **A $92,500 savings** a $92,500 savings here. These days, with the growth of satellite TV — two million earth stations were installed in the United

States by 1989 — there are more and more economical opportunities for guerrillas to use the persuasive powers of television to further their cause. And keep your mind open to direct-response television — two-minute commercials that make the offer and present a toll-free number by which credit-card holders may order. The average American household keeps a TV set on seven hours and seven minutes per day. TV sets are in 97 percent of U.S. homes, and the vast majority of them are color sets. Twelve percent of the country watches two or more home shopping programs. More than half of American homes now have two or more TV sets. No wonder TV is such a powerful weapon.

BILLBOARDS

Billboards are most effective if you can say these two magic words: "Next Exit." If you can't, billboards work best as a reminder of your other marketing. In most cases, if you want a billboard at a specific location, you'll probably have to rent several less desirable locations. Don't put more than six words **Six is the max** on a billboard if you want your message to penetrate the minds of passing motorists.

TELEMARKETING

If your offering is of interest to other businesses, telemarketing may be the most cost-effective marketing weapon in your arsenal. The main strength of telemarketing is its ability to get leads and close sales. The people making your calls should **Memorize the** actually memorize their script, but they should have it down **script** so pat that it does not sound memorized, and instead comes across as natural and spontaneous. Telemarketing experts report that if the response rate to your direct mailing is 2 percent, adding telemarketing will increase it anywhere from 6 to 22 percent.

Machines can handle your telemarketing now. They are capable of sensing that the person called has hung up, and they will disconnect in that case. If the listener stays on the line, the machine will deliver the sales presentation. If the listener wants to order, the machine will take all the ordering information. As effective as these devices are, there is talk of legislation to outlaw them as an invasion of privacy. Still, telemarketing is a versatile tool that makes an important addition to a guerrilla arsenal. You'll learn more about it in the next chapter.

DIRECT MAIL

Direct mail is so potent, so fast growing, so effective, and so important for any guerrilla to master that I've covered it in great detail in the next chapter, along with telemarketing and many other forms of direct marketing. Accomplished properly — and you'll learn how — direct mail can be the least expensive method, on a cost per sale basis, of all the marketing weapons available to guerrillas.

The Non-Media

Other important marketing weapons for guerrillas are the non-media, which include public relations, advertising specialties, free samples / demonstrations / consultations / seminars. Most smart marketers employ several non-media marketing methods to augment their standard media attack, while others employ *only* the non-media methods because they've learned that's what it takes for them to achieve success. This indicates (and I second the indication) that the non-media can produce maxi results, often at a mini cost.

PUBLIC RELATIONS

This non-media method often makes full use of the media in the form of free publicity in newspapers, on television, virtually anywhere. The basic strength of public relations, apart from its economy, is the credibility you get from a news story as opposed to an ad. Remember that people have built-in BS detectors when it comes to marketing. Public relations does not set off the alarm in the detector. The key element in gaining PR is news. The media hunger for news, so if your business is a source of news, there's a good chance you'll get the free publicity you seek.

PR gets past BS detectors

Another basic necessity for free publicity is a contact at the media from which you want the publicity, and the determination to contact that source until you get what you want. It's that lack of a publicity contact and a lack of time to prove determination that motivates many business people to hire a PR agent. These people charge up to several thousand dollars per month or per city to obtain news coverage for you. If they charge you, say, $5,000, then cause a front-page story on your company to appear in the *Wall Street Journal*, resulting in $50,000 worth of profits for your business, obviously they are a terrific value. A horrible yet common mistake is to obtain free publicity, but not have the product in full distribution or the service ready to go. Hold back on any publicity until you have everything all set — or you won't get the publicity again.

A common PR error

The very best PR is part of a marketing plan that also includes advertising. To run a profitable business on PR alone is a tough, almost impossible task. Don't even try. One of the easiest ways to obtain publicity is to offer your services as a radio or TV talk show guest. If you have something newsy, interesting, or fascinating for the people in your area, there's a good chance you'll receive an invitation to appear. In my area alone, radio guests on a major AM talk show in a recent month included a veterinarian, a home repair specialist, a pool contractor, a CPA, a plumber, a restaurant owner, a plant nursery owner, a cooking school teacher, an investment counselor, an art therapist, a midwife, and an auto garage owner.

ADVERTISING SPECIALTIES

Advertising specialties are the imprinted items that marketers give to prospects and customers. In the past, these were generally limited to calendars and ballpoint pens. These days, the items fill thousand-page catalogs. Call one or more of the advertising specialty firms listed in your local yellow pages and ask them to bring along a catalog when they visit you. You'll be dazzled. One of the three main strengths of advertising specialties is that they make the name of your firm more familiar to your target market, a worthy undertaking. A second strength is that advertising specialties enable you to advertise and/or mail to your target audience and offer them a *free gift* if they respond to your offer. A lot of the specialties make valuable free gifts that transform many a prospect into a paying customer. A third strength is that advertising specialties make friends for your business while unconsciously obligating those friends to become customers. At the moment, hot ad specialties include electronic gizmos, T-shirts, baseball caps, license plate frames, key chains, and videotapes. Along with offering specialty gifts in ads and direct mailings, consider handing them out at grand openings, special sales, anniversaries, and other noteworthy events. It is doubtful if even the best ad specialty can provide all the marketing firepower your firm needs, but as a guerrilla, you'll recognize the contribution it can make.

FREE SAMPLES, DEMONSTRATIONS, CONSULTATIONS, AND SEMINARS

The most sophisticated marketing company in the solar system is Procter & Gamble, as witness the remarkable success of their gigantic line of products. They are deeply committed to giving free samples of many of those products. This proves quality, earns goodwill, and gains confidence. P&G markets with free samples; smart guerrillas practice the principle whenever they can. If they can't give a sample, they give a demonstration. If that's not feasible, they offer a brief but free consultation. If they can't do that, they give a free seminar. All pay rich dividends when used properly. The primary strength of these freebies is that they give your prospects the closest experience they will have to owning your product or using your service. So if you offer high quality, a free shot of that quality will prove your point more effectively than any other marketing weapon. Free samples, generally prepared in minisizes, should be distributed to prospects only, and then, to as many as possible. A woman who sells $1 chocolate chip cookies at flea markets sells out every time because she offers tiny free samples of her baking wizardry. One teeny taste and the prospect is hooked. The sample cookies are so delicious, this method of marketing is almost illegal.

Free demonstrations prove beyond doubt the veracity of what you claim. If there is any way you can demonstrate benefits of your offering, you're well on the way to closing a sale. Ask any door-to-door vacuum cleaner salesperson. Don't worry about giving away valuable free information during a free consultation. Keep the time period down to thirty minutes, or make it one hour if you must — *no more* — then use the time to impress the pants off your prospect. Don't try to make a sales presentation. Instead, give the prospect honest and helpful information. The more valuable it is, the more likely it is that your prospect will want to become your client or customer. The idea is to show the tip of your entrepreneurial iceberg, making the prospect realize there's a lot more where that came from and it is worth the asking price.

Free demonstrations prove your points

Free seminars can be marketed with signs, circulars, and newspaper advertisements. During the seminars, which should also be no more than one hour, though some companies have achieved excellent results with longer seminars, give tons of crucial information during the first 75 percent of your seminar, then sell the attendees on buying your offering during the final 25 percent. If the first 75 percent was meaty enough, the final 25 percent will be amply rewarding. How many people who attend will buy what you're offering for sale? I've heard of closing rates as high as 67 percent for offerings (in this case, investing courses) selling for $595. I've heard of rates around 33 percent for reading courses selling for $395. But 10 percent is the norm. It's easy to see why so many people are now offering free seminars — which are really lengthy, live-action, high-involvement commercials. If you can do it, try it. And if you

An average of 10 percent will buy

can do it at your place of business, consider yourself fortunate. Telling people where you are and letting them have the experience of walking through your doors is a bonus for you.

Before embracing any of the free marketing weapons I've described and engraving them in your bronzed marketing plan, first you should test the idea. Then and only then should you make a commitment to employing these forms of marketing. For your sake, I hope you can use them because of their guerrilla-like nature: inexpensive and effective, requiring time, energy, and imagination from you. You can present them through your company or through the auspices of a local learning institution such as a university extension, a junior college, or an alternative educational source for adults. These institutes might pay you to deliver your seminar. You might offer to speak for clubs that would welcome your presentation. At these latter forums, you won't be allowed to sell, only to impress enough so that you can sell later. Think of the things you can give away, demonstrate, or inform about.

TRADE SHOWS, EXHIBITS, AND FAIRS

Single-weapon marketing

Alas, in a book stuffed with marketing weapons, I am compelled to admit that some of my clients have employed only one of the 100 marketing weapons and employed it to their delight and profitability. That weapon: exhibiting at trade shows. The major strength of trade shows is the buying mood present in most of the attendees. When people read your direct mail or see your TV spot or receive your sample, they may not be in a buying mood. But when they're at a trade show, they are there to decide what to buy, then to buy it. Get even more people to see your exhibit by distributing circulars about it — throughout the show, in the lobby, in hotels where attendees stay, even right outside the exhibit building. If those circulars are passed out by people clad in unique attire, tying in with your offering if possible, even more people will want to see what you're exhibiting. Recognize that in most cases, the main reason you are at the trade show is not really to exhibit your

offering, but to close sales. The most dazzling exhibit in the world is worthless if nobody is there to close sales to the people who are impressed by the goings-on.

Attendance at a trade show booth is intense, exciting, demanding, and exhausting. For this reason, have a crew of at least three or four attend to your booth. Consider having a party in your hotel suite. Invite your hottest prospects. Treat them like the royalty they wish to be treated like. Although a great percentage of business is accomplished on the trade show floor, don't underestimate the amount accomplished in the party suites. Trade shows are ideal venues for collecting names and distributing samples, brochures, and premium gifts. But they are more ideal for closing sales. Consider holding a sweepstakes at your trade show booth. Offer a truly valuable prize in a drawing open to anyone who puts his or her business card into the huge glass bowl at your exhibit. This way, you'll attract traffic, and just as important, you'll collect a galaxy of names for your mailing list.

Guerrillas are party animals

NEWSLETTERS, COLUMNS, BOOKS, AND COURSES

In case you haven't noticed, we're right in the middle of the Information Age. Newsletters, columns, books, and courses are part of that age as well as being part of a well-stocked guerrilla marketing arsenal. Such weaponry serves guerrillas well in the guerrilla marketing attack. The main benefit of these marketing weapons is that they establish your credibility — and credibility leads to confidence. These weapons also make you an authority on your topic — and authority leads to confidence.

With desktop publishing, using a computer, printer, and special software programs, it's easier, faster, and less expensive than ever to publish your own newsletter. Such a newsletter should give information rather than sell your offering, but still, readers should be asked and encouraged to buy what you are selling. The newsletter can be as short as two pages. And you can mail it every quarter, though every month is better. It's a

Be the authority

way of staying in touch, proving your expertise, giving beneficial information, and gaining confidence.

Writing a column or a book on a topic has been the key to many a guerrilla success. It must be a good book, clear and well written, but it need not be published by a big-name publisher. You can self-publish the book and attain almost the same esteem as if Houghton Mifflin had published it. Then use it as part of your marketing arsenal. Take it from one whose marketing business has benefited from books he's written: A book is a potent weapon. Teaching a course is, too.

In addition to these obvious ways of establishing your authority are giving speeches, creating an audiotape, producing a videotape, writing an article for a local or national publication, and appearing as an authority on a local TV or radio show. These non-media marketing methods are part of the guerrilla marketing attack. And for many, they're a lot of fun.

THE COMMUNITY

One of the most valuable yet overlooked marketing avenues is your very own community. Opportunities to market are abundant, and costs are minuscule. The greatest strength of utilizing community marketing opportunities is that they lead to community involvement and make you a community presence. To a guerrilla, that's a very desirable thing. People want to buy from friends. When you are part of the community, you become a friend. Marketing through the community affords you limitless opportunities for tie-ins. For example, you can tie in with businesses in your community. Businesses are generally very receptive to tie-ins, but it takes a guerrilla to instigate the arrangement. Connect up with your community by offering special discounts to members of a large business located near yours, and distribute special discount cards to all employees. Schools in your community represent excellent tie-in possibilities. Perhaps you can hold a demonstration at the school or you can contribute goods or services for one of their school fund-raising events. Schools include day-care centers, nursery schools, grammar schools, middle schools, high schools, private schools, religious schools, junior colleges, community colleges, trade schools, even a nearby university.

Scour your community for local charities, then help them with goods, products, time, energy, and methods of raising funds for their usually noble work. This helps you and helps others at the same time. Everyone wins. Market through **Everyone wins** churches, again helping them raise funds, advertising in their local paper, obtaining write-ups in their newsletter, posting signs on their bulletin boards. These days, there is a plethora of clubs with which you can tie in. Put up your circular on the bulletin boards of health clubs, bridge clubs, recreational clubs, running clubs, bicycle clubs, ski clubs, service organizations, tennis clubs, golf clubs, and any other groups to which your prospects might belong. Regularly, drive through your community, looking for opportunities to help it, to tie in with existing entities, to become an integral part of it. Think of these drives as guerrilla scouting patrols.

MISCELLANEOUS MINI-MEDIA

Every day, a new marketing medium is developed. Guerrillas check into all of them because some can break the bank. Some of the newer marketing media include:

- Ads in building restrooms, including all stalls
- Ads on parking meters
- Positive picketers who picket your business with signs that say how good your business is; if you employ this tactic, call the newspaper and they'll probably cover it. Call the local TV station, too.
- Inflatable, helium-filled objects of any kind
- Ads on videocassette boxes
- Ads on trucks and truck tops
- Ads on bus shelters
- Ads on shopping carts
- Postcard decks
- Marquee-type messages on blimps

A comprehensive knowledge of the media possibilities is important intelligence in any guerrilla marketing attack. The times are favoring guerrillas by making so many media available to them, such as newspaper zone editions, magazine regional editions, cable TV, satellite TV, bingo cards in magazines, newsletters, and inserts. The influx of bulletin boards should attract many a sign and circular from a victory-minded guerrilla. And wait till you see the growth of videotape marketing, which offers a free brochure in the form of a brief videotape. It's just in its infancy now, but companies trying it report returns of 17 to 20 percent for each videotape they give away for free. The tapes should be given free rather than loaned so that they may be viewed more than once and shown to others.

You now know the breakthroughs in psychology that can benefit you as a guerrilla. And you know which media will work best for you to put your new understandings of human behavior to work. But before launching your attack, let's take a closer look at direct marketing.

9
Direct Marketing, Guerrilla Style

IN THE 1970s, fully 50 percent of Americans never bought anything as a result of direct marketing. But in 1987, over 90 percent of Americans bought at least one thing purely as a result of direct marketing. Americans have learned to trust direct marketers. They love the convenience, the guarantees, and the vast number of offerings available by phone or mail. As we move toward the year 2000, direct marketing will become even more important, especially with the coming growth in interactive TV — seen now in the influx of TV shopping shows.

Although telemarketing is now the primary method of direct marketing, and television is moving up fast, direct mail will always be a major marketing weapon for small and medium-sized businesses because it is so inexpensive and easy to test. A wealth of specific advice about creating direct marketing can be found in *Guerrilla Marketing*.

As a practicing guerrilla, you should realize that direct marketing refers to:

- Direct mail
- Couponing in newspapers and magazines
- Postcard decks
- Direct-response television
- Direct-response radio
- Telemarketing
- Matchbooks
- Electronic bulletin boards
- Inserts

- Statement stuffers
- Catalogs

As you can see, direct marketing is any marketing that is intended to produce profits without a face-to-face meeting. These days, about two thirds of all direct marketing is aimed at individuals, while one third is aimed at companies or institutions. The volume of direct mail alone is in the range of 50 *billion* pieces per year.

In 1987, a study was undertaken to determine which marketing method worked best for retailers. A majority, 46.6 percent, said direct mail was the most effective; 32.4 percent said newspaper advertising was best; 9.1 percent rated TV number one; 4 percent voted for radio; 7.9 percent had no idea.

The direct-mail explosion has been resounding since the 1970s and continues to increase at a rate of more than 10 percent per year. In 1982, telemarketing expenditures overtook direct-mail expenditures. Yet the fastest growing segment in the whirlwind world of direct marketing is now postcard decks. Stay tuned.

Direct marketing is hardly new. It has been used with great success by the International Correspondence Schools since 1891. Today, they continue to use it, as well they should, since they receive around 800,000 inquiries yearly, many from coupons in books of matches, as a result of their commitment to this marketing.

What are the reasons for the surge in direct marketing?

1. It creates a sense of urgency. When guerrillas use direct marketing, they always alert the recipient to the date that the offer expires. Knowing there is a deadline motivates many people to take action rather than to think it over.

Accountable and honest

2. It is accountable. Make an offer with any direct-marketing vehicle and give it a cutoff date of, say, March 15. By March 16, you know whether your effort soared or fizzled. Standard media marketing does not give such instant and honest feedback.

3. It is economical. More and more small businesses are realizing that regardless of the overall cost, direct mar-

keting enables them to market their offerings at the least **Low cost per** expensive cost per sale of any way they have ever tried. **sale** They're learning it is far more economical than face-to-face sales calls, which are estimated to run around $275 each. No matter how expensive the postage, the brochure, the letter, or the contents of the mailing package, they're going to run considerably less than $275.

4. You can determine the exact cost to you for each sale you make, each dollar of profit you earn. A true guerrilla understands the mathematics of direct-response marketing. **Understanding** Anyone who engages in direct marketing without knowing **the math** with certainty the exact cost to acquire a new customer, the lifetime worth of that customer, and the profit per sale is only playing with marketing, not taking it seriously. Marketing is too expensive to be treated capriciously.

5. It is a very convenient method of ordering merchandise. From the comfort of one's home or office, people can consider offers, then accept or reject them. If they do accept them, they can receive what they order within days, though most people don't mind waiting weeks. Guerrillas give their prospects the option of selecting standard delivery time or quick delivery time at an added cost.

6. Businesses of all sizes are discovering that direct marketing is the most effective marketing weapon of all if they lack an established means of distribution or if they have a poor location.

7. Direct marketing offers the ultimate in selectivity. You can choose the people to whom you will market directly on the basis of their sex, age, occupation, education, special interest, religion, marital status, geographic location, and past history with other direct marketing offers. You can choose the businesses to which you will market based on their probability of accepting your offer. You can also market to your past customers only. If you're a guerrilla, you will — again and again.

8. You can personalize the marketing vehicle to deeply involve each prospect and prove that you recognize his or her special nature. You can also provide enough infor-

mation for the prospect to make an informed buying decision, and you can appeal to both left- and right-brained people.

9. You can test your direct-marketing vehicles and offers at a relatively low cost. Once you have established proven vehicles and offers, you can use them to reap enormous profits. This testing enables you to be wildly experimental, utilizing mailing lists, phone scripts, and offers of all kinds. You may end up with response rates ranging from .00001 percent or lower to 87 percent or higher. I've seen both, and I don't have to remind you that five miserable response rates are a small price to pay for one glorious response rate. You can use the elements of the glorious test and employ them throughout your marketing area, regardless of its size. That's better than money in the bank, because it can multiply your profits with certainty.

One is enough

Why direct marketing is growing

In addition to these advantages, direct marketing can help you support distributors and dealers, reach buyers from small towns not visited by salespeople, announce new products and price changes, convert sales calls into sales, obtain hot leads for salespeople, supplement your other forms of marketing, develop new markets, build up marginal territories, and increase your profits by eliminating middle people. No wonder direct marketing is growing at such a healthy pace!

As recently as 1970, direct marketing meant sending a letter. But today, to a guerrilla, direct marketing means sending a letter, then a follow-up postcard, then one more letter, then a follow-up phone call. The days of obtaining the order with a single effort, while not gone forever, are rapidly diminishing. To succeed, think in terms of a direct-marketing *campaign* rather than a direct-marketing piece.

Succeed also with *postcard decks*, the very latest in direct-mail devices. A postcard deck is a package of 15 to 25 postcard coupons wrapped in transparent plastic and directed to people of certain demographic groups, such as investors, affluent people, psychologists, doctors, and the self-employed. The types of groups increase each week as these decks, generally containing special offers, new items of interest, free information, and even gifts, earn profits for those who use them as well as those who publish them. High-energy guerrillas use them as both a marketing weapon and a profit center — combining a few postcards for their own enterprise along with many postcards for compatible enterprises. Naturally, the compatible enterprises pay for the privilege of being included in the postcard deck.

Your competition in direct marketing comes from all directions. Absolutely everyone who mails an offer to your prospect is the competition. Everyone who runs a coupon, inserts a preprint into the newspaper, calls a prospect, runs a TV spot with a toll-free 800 number is the competition. There's an army of competitors lurking out there, battling to win your prospects' dollars.

To be a guerrilla, learn from other guerrillas. That means **Learn from other** you should not only read this chapter and put its principles **guerrillas** into practice, it also means you should get on mailing lists yourself to see how great direct marketing captures your attention. Don't copy those ideas, but let them inspire you. Of every twenty pieces of direct mail that come your way, be on the lookout for the one or two that cry out to be opened, that tempt you with enticing offers.

When creating or judging your own direct marketing, be sure you practice the basic principles of repetition. Put your prime idea right up front — not buried somewhere in the body of your letter, brochure, or postcard.

Although your direct marketing may be addressed to thousands or even millions of people, remember that they will read your message one person at a time. Write it with this in mind. Act as though your mailing list has but one person on it, and work like crazy to get that person's order.

What are the goals of direct marketing?

In direct-mail parlance, a "classic" mailing package consists of an outer envelope, a letter, a brochure, a response device (coupon or phone number), and a return envelope. When using direct mail, you always have three goals:

1. To get your letter opened
2. To get it read
3. To get the order

Obviously, you can't succeed at the second two goals if you don't succeed at the first. One way to succeed at the first is to write a "teaser" on the outer envelope — a brief set of words so fascinating that the recipient just must open the envelope to learn more. Another way is to make your offer via postcard. Then you don't even have to worry about the first goal.

How do you create effective direct mail?

Be sure that your mailing is centered on a clear and powerful *idea*, an idea centered on an offer to your prospect. Relate everything in your mailing to that idea. Tell it at least three times — at the beginning, in the middle, and at the end.

Be long yet short   No matter how long your letter, use short words, short sentences, and short paragraphs. Use subheads liberally. Even though the mailing piece may be long, it can "feel" short. An eight-page letter can be more readable than a two-pager. Remember that long copy works better than short copy. An old and true marketing adage is "The more you tell, the more you sell." Even though it rhymes, it's true. Of all the things people dislike about marketing, "lack of information" comes in second. (A "feeling of deception" comes in first — that's why people have those built-in BS detectors I mentioned earlier.)

Keep in mind that people read the first line of a letter first, the P.S. and signature second, and the rest of the first paragraph third. Fourth come the subheads. After that, if they are enticed, they'll read the rest of the letter. Maybe even the brochure. And if you've done your work right, the response device, too. Guerrillas not only tell their great idea three times in both the letter and the brochure, but they also clearly state it in the response device. You'll find many specific guidelines for creating winning direct-mail packages in *Guerrilla Marketing*.

The sharpest guerrillas write the response device first. They do this so that they are clear on what the reader is supposed to do. This way, their letter and brochure can increase the momentum and relate as closely as possible to the response they wish to elicit. This is called "working backward" and it's the smart way to work. Many experts feel that the response device is the most important piece of the package and frequently the most poorly conceived. Be sure it restates your offer briefly but completely. **Working backward**

In your response device, letter, and brochure, state your offer in a manner that attracts the attention of your prospects, interests them, offers them a benefit, creates a desire within them for your offering, and calls for action by a specific date. Tell them the exact action they must take: call a toll-free number, complete and mail a coupon, come in to your place of business, whatever you wish them to do.

Be sure not to write incomplete letters or brochures. Some people read letters, others brochures. Few read both. So both the letter and the brochure must be self-contained, give the details, make the offer, repeat the offer, and ask for an order.

And don't secure a reader's attention with one thing, then switch subjects mid-letter. I call this the "Speaking of insects, how's your aunt?" school of marketing. People resent being tricked into reading something. If you don't think they'll be interested in your offer, don't mail to them in the first place. Guerrillas are never sneaky.

Direct mail at its simplest is a standard-size postcard; direct mail at its most complex is a several-hundred-page full-color catalog. Naturally, the idea is to start simple, then build up to more complex mailings as long as they produce the results you want. Some mailings are one-shots designed to promote a sale, introduce a line, or announce a new offering. Others are sequential mailings — four, five, or six mailings all repeating the same or a similar offer. You can expect a decreasing response with each mailing. A third type of mailing is a continuing series of mailings, regularly scheduled and aimed at the same mailing list, but making a different offer each time.

How important is testing?

Talk to any direct-marketing expert and you'll learn that testing is crucially important. The six rules for successful testing are:

1. Test only one thing at a time, and test it against a known standard, called a "control."
2. Know what to test and what not to test.
3. Obtain enough responses for statistical significance. Anything under twenty is not enough.
4. Keep meticulous records of everything you do, and be sure to save at least one dozen samples of each mailing.
5. Never change anything that testing hasn't shown that you ought to change.
6. Always be testing something in each mailing. Guerrilla marketers are notorious experimenters.

When you are testing prices, test major differences in price, not minor differences. Test a $129 price point against a $99 price and a $149 price, not against a $119 or $139 price.

To forecast your results, if you send a mailing first class, figure on double the response you've received at the end of two weeks' time from the date of your first order. If the mailing went third class, double the response you've received at the end of four weeks' time.

Frequently, a company will want to test a price, a mailing list, and a free premium for ordering. A guerrilla knows that this requires three tests. A novice thinks all three can be tested at once. So if the mailing is a failure, the novice doesn't know whether to blame the price, the list, or the premium.

Guerrillas don't waste their time and money testing little things. Instead, they test important elements such as offers, prices, mailing package configuration (classic vs. self-mailer vs. oversize envelope vs. anything else), times to mail, mailing lists, and the presence or absence of a toll-free number. These are the arenas where success or failure is determined.

How important is the right mailing list?

Actually, it's not much more important than air or water — that is, it's crucial for success on the direct-marketing front. The experts tell us that the keys to any successful direct-marketing effort are:

1. The right mailing list
2. The right offer
3. The right statement of that offer
4. The right financial formula

That sounds easy, but it's not. The best mailing list of all is your own customer mailing list, that list you should have been compiling since the very first day you operated your business. It is a list that should grow regularly, that should be mailed to regularly, that should be updated regularly. Be certain it is kept up to date. That means the addresses should be correct and the customers should still be interested in your business. If you think there is a direct correlation between the size and timeliness of your customer mailing list and the size of your bank deposits, you're thinking like a guerrilla.

Compile that customer mailing list by keeping track of your customers' names and addresses via receipts and order forms, by asking customers to sign a book that is located near your cash register, and by holding sweepstakes, as I discussed earlier in the book, that entice people into your place of business in order to sign up. That latter group will not quite be customers, but at least they'll be familiar with your business. Familiarity leads to confidence, and we all know the importance of confidence.

A great offer, magnificently stated, and mailed to the wrong list is going to be a major disappointment to you. So be sure, when selecting your mailing list, that you are focusing on people who are likely to want your product or service. You can get a lot of help in selecting lists from a mailing list broker. Find these people under "mailing lists" in your yellow pages. Tell them your target audience and enlist their aid in finding lists of people that closely match up with that audience. Don't try to save money with a cut-rate mailing list. The list broker you consult will have countless lists; that same person will have clear recommendations as to which have the greatest probability of generating profits for you. The purpose of direct marketing is not to save money, but to create profits. Never forget that.

Appealing offers   The types of offers that will appeal to the people on your list are offers of free gifts with a purchase, offers of substantial discounts, offers of new or unique products or services, and offers of proven interest to your list. For example, if you're going to have a sale, mail to your customers first, letting them know they have first crack at the merchandise because of their "preferred" status.

After you have selected your list and offer and have proven **Direct-mail** both in a test mailing, measure your actual costs against your **expenses** projected profits. Those costs will be for postage, paper, printing, typesetting, artwork, copywriting, preparation time by you and your employees, the mailing list, and delivery costs. Be sure to include all your costs. Some are fixed costs, such as postage and paper; some are amortized costs — one-time charges for items such as copy, artwork, and photography. These costs will help you get a fix on your break-even point. Not knowing that number excludes you from the ranks of pure guerrillas and potential profit makers. If you can't turn a profit after several tests and solid consultation by pros, perhaps direct mail isn't for you. But for your sake, I hope it is.

Response rates vary from business to business. For some offerings, such as educational seminars, a good response rate is one half of one percent of the total number of pieces mailed. For other offerings, such as a two-for-one meal offer, a good response is between 10 and 25 percent. Once, a bank retained me to create a mailing package for their automated teller machine card. The response rate was 87 percent. Another time,

I prepared a mailing offering a book on free-lancing to free-lancers. That time the response rate was 34 percent. If you achieve good results with your own test, don't wait to roll it out. If you mail to 5000 names from a list of 100,000 names, and you obtain a profitable return, mail to those other 95,000 people ASAP. Times change, and these days they change faster than ever.

Learn five basic direct-marketing skills.

To get in on the money to be made through direct marketing right now, start developing five basic skills:

1. The skill to develop fresh, new marketing ideas. A treasure trove of those ideas is upcoming in a moment.
2. The skill to write captivating opening lines. Recognize the crucial importance of the first line of copy, indeed, the whole first paragraph. If they don't invite your readers to read on, you've probably lost the sale. A prospect might read at least a single paragraph. After that, you're on your own. It's all up to the offer and ideas expressed in paragraph one.
3. The skill to harness the power of a potent P.S. Remember that it is almost as widely read as the opening line. Make your P.S. a restatement of your offer, your prime benefit, your main idea, a request to order right now, or even something esoteric that requires reading the rest of the letter for details. But whatever you do, don't leave out a P.S. for any reason.
4. The skill to state your thoughts with brevity. That doesn't mean a brief letter, but brief elements within it.
5. The skill to be different *and* better, not just different. Although guerrillas might test postcards, new media, unique packages, and other direct-marketing innovations, they love the economy and track record of the classic package.

Guerrillas do not apply standard advertising techniques to direct mail. You can be sure they shun clever headlines, uptown artwork, image-building graphics, and short copy. They realize that few advertising agencies have direct-marketing skills. While I was an ad agency hotshot, I certainly was oblivious to the talents necessary to succeed at direct marketing. And I was with the largest ad agency in the world! You can be sure I was not alone in my ignorance. Standard ad techniques seem shallow and tinny when applied to direct marketing.

Never stop testing

Guerrillas don't stop testing — *ever*. Sure, they're elated at a record-breaking response rate that sets new highs for sales and profits. But they figure records are made to be broken, so they keep on testing. They test new offers, copy, artwork, package components, mailing lists, products, services, and direct-marketing tools. One of the world's most prolific advertisers, Procter & Gamble, never stops testing new TV commercials for its products. This goes to show that *some* standard advertising practices are compatible with direct-marketing practices.

In all direct marketing, be sure you stress your key ideas. Do it with underlining, putting brackets around a key paragraph, using a special color emphasizing your key points, with a yellow high-lighter effect, or with all capital letters. Guerrillas give their target audience all the help they can. Stressing key points is one way to help. But don't stress too many points or you'll send your readers scurrying to the medicine cabinet for an aspirin.

Learn fifteen smart direct-marketing ideas all guerrillas know.

1. Although direct marketing can work during all twelve months, the three best months are January, February, and October. September comes in fourth by a whisker. All four of those months signify the beginning of something: the new year, the coming of spring, the winter season, and the school year, respectively. March, April, and November are generally 77 percent as good as the three best months.

May, June, August, and December come in at from 50 percent to 75 percent of the good months. July and August are traditionally dismal months for direct marketing — probably because people are more involved with outdoor activities. Guerrilla rule of thumb: The worse the weather, the better the climate for direct marketing.

2. Think creatively about stamps. Use multiple stamps, as I discussed earlier, a commemorative stamp, or a foreign stamp mailed from a foreign land. This is an area for imagination rather than big bucks, an ideal battleground for guerrillas.

3. Avoid the use of cheap envelope paper, flimsy letter paper, screaming type, and anything else that cries out "junk mail!" It amazes me when I meet clients with powerful offers, willing to pay me a fancy sum for writing the mailing package, willing to pay top dollar for mailing lists, but unwilling to put it all together on expensive paper stock. The pennies they save cost them many dollars in profits. A mailing package is only as strong as its weakest component. Words to remember, if you're a guerrilla.

4. Monarch-sized envelopes (3 ⅞" × 7 ½") work better for some mailings than standard #10 envelopes. That makes them worth testing. They also can offer economy as a bonus to you. White and off-white envelopes are the most likely to be opened. Color-loving experimenters have conducted tests of every color you can name. White beat them all.

5. Indent your paragraphs to give your letter a more personal **Be intimate** feeling. Don't forget, people will read your letter one at a time, so capitalize on the intimacy.

6. Use your own letterhead when mailing to your own customers. They're actually happy to hear from you and will place a high priority on your envelope and its contents.

7. Saying "Personal and Confidential" will increase the number of people who open your envelope. But what you say inside had better be personal and confidential or you will have alienated a prospect.

8. Even though computers make it easy to have an even (or

"justified") margin on the right side of your letter, don't succumb to the temptation. Your Aunt Maude never used justified margins, and you always enjoyed her letters. Your readers will be put off by letters that appear typeset and impersonal.

9. Many tests show that letters with photographs showing your product in use often perform better than letters without photos. In my own experience, letters with a color photo of the "free gift" the recipient will receive for responding perform best of all. And when that color photo shows through the window envelope, along with the line "A free gift for you!" — gangbusters!

10. The best color combination for attaining a healthy response is: white paper, black type, and blue signature, underlines, and/or margin notes. I've seen a red headline printed at the top improve the response rate even more, but that red was used because the letter referred to a Valentine's Day offer.

11. In a brochure, use drawings if you must, but realize that photographs improve response rates. Photos are more believable, more intriguing, and grab a reader's interest more than any drawing. Hint: Whichever you use, be sure to include a caption. Captions get extremely high readership.

Toll-free numbers
12. Toll-free numbers increase direct-marketing response rates, especially mailings to individuals. If you don't have a toll-free number, invite collect calls. Surprisingly, if you mail to people in an organization, most of them won't call collect when they do call.

13. Print copy against white, never against black. Test after test proves this, yet art directors love reversing type (white type against a black background). The power of art directors is not to be underestimated. They are talented, persuasive, and often misguided when it comes to direct-response marketing. Direct marketing that is over–art directed might win awards, even get hung on the wall, but it doesn't pull in orders. Seek out art directors who specialize in direct marketing. They are few and valuable.

14. Use postcard mailings whenever possible because postcards

don't have to be opened. Postcards can be oversized (6″ × 9″), can have long copy, can have a response device attached, and can do much that a classic package can do. But their best feature is that they are already opened and ready to read. That means your foot is in the door, and the momentum leading to the sale has already started.

15. Present your mailing package in the form of a survey. Actually include a questionnaire along with the rest of your classic package.

Guerrillas who run new and small companies can do many things that larger companies can't. These include staggering their mailings, mailing only a few letters each day or week. They can also type the envelope and never use a label. They can sign their own name and even handwrite their P.S. I don't recommend handwriting the address on the envelope because it lacks the professionalism of a typewritten address. But the other things do add a personal touch.

To learn the most possible about direct marketing, more than is available in any book, I recommend that you subscribe to *Direct Marketing*. Get it from Hoke Publications, 224 7th Street, Garden City, NY 11535. I'm surprised they don't have a toll-free phone number.

What every guerrilla should know about telemarketing

As hundreds of thousands of thriving businesses know, the telephone is one of the most lethal of all guerrilla marketing weapons. Telemarketing, along with direct-response television, is one of the most rapidly growing areas in direct marketing. Telemarketing is growing even faster than direct mail, which continues to grow at breakneck speed. Two thirds of telemarketing is business-to-business, while one third is business-to-consumer. That's primarily because four times as many industrial firms as consumer firms market by telephone. And that makes sense because the average industrial order is between $800 and $1000, while the average consumer order is in the $50 range.

The components of telemarketing

Telemarketing has three components: the preparation, the call, and the follow-up.

The preparation should eventually net you not only a list of prospects and the best time to call them, but also an actual telemarketing script. The person using it should memorize it to the point that it sounds conversational. He or she should sound friendly, ask questions, listen, speak at a sensible rate of speed (not too fast; not too slow), and use the prospect's name. The idea still is to close sales, although telemarketing is also used to upgrade an order, make an appointment, arrange for a demonstration, secure referrals, and engage in market research. You can obtain more details and take a look at a sample telemarketing script in *Guerrilla Marketing*.

As with face-to-face selling, 30 percent of people called will be easy to sell, 30 percent will be impossible to sell, and 40 percent will show you who's a salesperson and who's an order taker. You don't want any order takers clogging your phone lines. Salespeople are required if you're to succeed at telemarketing. That 40 percent who may buy or may not buy, depending on the quality of the teleseller, will probably influence the profits of your business as much as or more than any other factor.

AT&T tells us that there is stress in telemarketing, but that it is reduced if you call qualified leads, people who have requested information, or those with whom your company already has a relationship. Many of the direct-mail rules apply to telemarketing, especially the part about constantly testing and improving. It takes about a hundred calls to conduct a proper test. Telemarketing success dramatically improves if the person to be called first receives a letter and brochure stating

When to call

that you'll be calling. The call should come the day after the letter arrives. It improves more if a letter is sent right after the call is made, confirming all decisions. The whole idea is that telemarketing should be an important part, but not all, of a direct-marketing program.

Once you've got a prospect on the line — and it takes six attempts to accomplish this — you've got 45 seconds to stimulate enough interest for the person to want to continue the conversation. Because there is so much rejection in telemarketing, make sure none of your people do it more than four hours a day, including breaks.

Success in direct marketing should be measured by only one gauge: profitability.

Direct marketing through magazines and newspapers, using coupons and/or toll-free numbers, draws tiny response rates, but the population from which it draws can be so large that response rates aren't nearly as important as simple profitability. If you invested $2,000 in magazine and newspaper advertising, and enjoy $22,000 in profit, what do you care if your response rate was only .0001 percent?

All direct marketing lives or dies by the quality of the offer. From the offer spring the profits. Successful mass media direct marketing also depends a great deal on your headline. Photos work better than drawings — be sure to include a caption — and don't fear writing long copy. Readership falls off after fifty words, but between fifty and five hundred words, there is little loss of readership. Use your coupons and toll-free numbers to get orders or requests for brochures.

The omnipotent offer

Television as a profitable direct-response medium is growing, too. I'm sure you've seen myriad offers for phonograph records, not to mention kitchen appliances and more phonograph records. Many of those direct-response TV spots are two minutes long. The best ones devote a minimum of twenty to thirty seconds giving ordering information: price, toll-free number, address, credit cards accepted. In addition, many spots are purchased on a per-order basis, with the TV station being compensated for each order you receive. That's fair for both of you — if you receive enough orders.

Two-minute TV shows

My best advice about direct-response TV is to create it so

that it does not look like direct-response TV. My congratulations go to Time, Inc., which uses direct-response TV with creativity, verve, and tempting bribes. Their hard-hitting direct-response TV commercials are as interesting as many programs and very unlike the usual hard-sell, visually unappealing direct-response TV efforts.

A lovely part of this medium is that it costs so little to test in a small metropolitan area. Often, the TV studio will give you a low rate for producing your commercial — don't forget, you should write it, not them — if you're going to run it on their channel. The possibility is worth checking, because TV reaches so many people.

As with other mass media direct marketing, the TV direct-response rates are tiny, but the profits are often gigantic. Such vast numbers of people watch television that it doesn't take a high percentage of responses to make your offer a resounding success. Clients of mine report a better response from late-night commercials than from prime-time commercials. That's because the late-nighters are conditioned to expect TV offers.

The informercial   A new and unique opportunity for many guerrilla-minded companies is the coming of the *informercial* — the thirty-minute TV show that's really one long commercial. An informercial contains twenty-four minutes of programming material, always related closely to the product or service offered, plus six minutes of commercial time. These are best utilized with three two-minute commercials. One of the best things about informercials, which are available through satellite networks, cable stations, or directly from TV stations, is their low cost. Today, informercials are available for as little as $200. Of course, what this buys you is half an hour of nonprime time, but a whole lot of people will be watching. And if you fill the half hour with fascinating visuals, dynamic information, and a tempting offer, many of them will be buying. Look for this guerrilla marketing opportunity to grow with lightning speed as we move toward the twenty-first century and TV options grow as costs plummet.

If you're to be a guerrilla, explore direct marketing, investigating the many ways it can help you. Do a direct mailing.

Try telemarketing. Go for direct response with a newspaper coupon. Offer a product or at least a free brochure in magazines. Give a shot to TV if it is appropriate for your product. Seriously consider an informercial for your offering. Learn more about them from any good media-buying service. The best I've encountered is CPM, 240 E. Ontario, Chicago, IL 60611, (312) 440-5200. The low cost of testing makes direct marketing very attractive to guerrillas. They know that a test which fails costs very little and teaches a lesson. And a test that succeeds can lead to handsome profits.

Expect failure with many of your tests; expect success with few. But those few will finance those many with ease. Just remember that all you need is for only one test to succeed. Ideally, you'll be able to ride that horse across the finish line and end up a winner.

10
The Most Important Person

IF YOU DON'T UNDERSTAND that the most important person in the guerrilla marketing attack is unquestionably *your customer*, I urge you to put someone else in charge of your marketing. The guerrilla marketing attack centers on the customer, reveres the customer, lavishes service, quality, and energy on customer satisfaction.

As more and more businesses learn that customer satisfaction is absolutely essential to long-term success in the marketplace, more and more people are learning to expect and demand first-rate service. Yet, most businesses are woefully unaware of the need for conscious customer devotion, an attitude that must be present in every level of your company, in the heart of every employee, and at all times. A tall order, indeed, but not for a guerrilla — because a guerrilla realizes that customer satisfaction begins with a customer's first contact with you and *continues for as long as your company is in business*. Eighty percent of lost business is the fault not of poor quality but of failure to continue in the effort to maintain and build on customer satisfaction.

The U.S. government offers no Index of Customer Satisfaction. But your customers and prospects know the feeling of customer satisfaction. And they want it back again. Thomas Peters, coauthor of *In Search of Excellence*, says, "In general, service in America stinks." This is generally blamed on the inflation of the 1970s, which forced many businesses to eliminate many services in order to keep their prices reasonable Deregulation spawned more price wars and more service cutbacks. Companies discovered that they could run themselves with self-service and with computers. No guerrilla would take heart in this discovery. A guerrilla would realize that the human element would be too conspicuously absent, and this element breeds far more customer loyalty than any computer.

The humans in your operation should be well trained in customer service. L. L. Bean, the highly successful mail order company, requires that each employee undergo forty hours of training before dealing with the public. All five thousand cab drivers in Miami, Florida, are now required to take a three-hour course in courtesy (the course is called "Miami Nice"), which has led to an 80 percent reduction in the rate of customer complaints. Take a hint from these examples. And heed the words of Karl Albrecht and Ron Zemke, authors of *America! Doing Business in the New Economy*, when they say, "Service people can become so robotized in their actions that they greet any customer request with a standardized response." That may explain the Department of Transportation's announcement in 1986 that complaints about poor airline service increased 30 percent that year.

Does your company have a customer recourse policy — a method of dealing with dissatisfied customers? If not, here's a good policy to use as a start:

THE CUSTOMER IS ALWAYS DEAD RIGHT EVEN WHEN THE CUSTOMER IS DEAD WRONG.

Don't stray too far from that policy, and you'll be able to convert disgruntled customers into repeat customers who gladly tell others of the superlative service they received from you.

Am I actually saying that a dissatisfied customer is a guerrilla marketing weapon? That's exactly what I'm saying. And I'm saying that a satisfied customer is also a guerrilla marketing weapon. I'm repeating in a variety of ways that every customer is not only a marketing weapon but also a human being of great importance to your life and your business.

First, you must attract your customers. You do that by practicing the principles you encountered in the first nine chapters. But if that's all you do, your business growth curve will soon be headed in the wrong direction. The nonguerrilla temporarily glows with smugness and self-satisfaction after making the sale.

The nonguerrilla is wooed and won over by the fast buck, the instant gratification, the quick fix. As a guerrilla, you, too, should justifiably feel the thrill of having attracted your customers. You should possess the wisdom that the attraction was only the first step. Few businesses thrive if they take only the first step.

The dance begins The dance has just begun. There are more steps to take. One step may be thanking the customer by mail or phone. Another may be writing to this esteemed personage and making certain that all is going smoothly, asking if you can do *anything* to make the customer happier after having purchased your offering. And one more step might be offering the customer a special bargain on a related service or product. Once you prove that you render caring service, you can ask the customer to tell his or her family, friends, and neighbors about your business, reassuring the customer that you will bestow a generous reward, in the form of a gift of value to the customer, if any new customer merely mentions the old customer's name.

The dance continues. You send the customer mailings that offer values, discounts, new service and product information, data that you know will be useful to this valuable person. You invite the customer to special sales or seminars or parties or unveilings or demonstrations or any type of event that intensifies the customer's relationship with you. You make the customer a member of your "club." Perhaps you send the customer a gift, such as a great-looking coffee mug personalized with your company's name *and* the customer's name.

At all times, it is crucial that you remember that the customer is an individual and a human being who is *never to be taken for granted*. Because of this truth, overlooked by many businesses, you'll be certain never to render impersonal service.

The worth of a single customer cannot be overestimated. Here is an example to prove the point. The business owner who is oblivious to the nontraditional methods of marketing will sell a customer a $100 item, and take pleasure at the $50 profit just earned. "Hot dog!" thinks the nonguerrilla. "That customer was worth fifty dollars to me!"

Does the guerrilla business owner think the same way? No,

the guerrilla sees it quite differently. The sale of the $100 item is made, earning a $50 profit for the guerrilla. But the guerrilla realizes that with proper service and conscientious customer follow-up, that customer can be worth $100,000 over the life of the business. $100,000???!!! Yes, $100,000. Maybe even more than $100,000. **The $100,000 customer**

The guerrilla marketing attack requires solid follow-up for repeat sales that can generate *four* $50 profits during the course of the year, a total of $200 in profits that first year. More follow-up will result in a continued relationship over the lifetime of the business, another twenty years. Twenty years times a $200 profit is a $4000 profit from one customer.

Sustained, caring follow-up will result in referrals and word of mouth that spawn a minimum of five more customers. That represents a potential profit of *$20,000 more.* And if those five customers receive first-class follow-up, they might spread the word to five more, *driving the profit from that first customer up to $100,000.* And higher, *much* higher.

That's why guerrillas act so lovingly and energetically *after* the sale. And because so few other business owners act this same way, they lose most of their business because of an uncaring attitude after the sale.

Apathy after the sale is unknown to guerrilla marketers. It is a fatal disease that they avoid with the same care they exercise in avoiding smallpox and rattlesnakes. True guerrillas know the value of customers deep in their bones. And they know this feeling is not nearly enough. They realize they have an obligation to convey the concept of what I term "customer love" to every single one of their employees. Sure, their management team may be the first to know. And their salespeople certainly know, as do their service people. So do their delivery people, their maintenance crew, their billing department, their materials handling crew. Everybody is fully aware of the company's steadfast dedication to the customer. This orientation to service and follow-up is so pervasive the customer actually feels it. Guerrilla marketers make sure that their customers feel it, read about it, see it, hear it, even taste and smell it if possible. **Customer love**

Everybody knows that even though the boss signs the paycheck, the customer is the one who comes up with the cash. Therefore, in addition to the primary goal of your business, one more goal, if you're to be a guerrilla marketer, is total devotion to your current customers — matched only by your devotion to enlarging their numbers. There will be a clear and joyful correlation between the size of your bank account and the size of your customer list. Customers are to be attracted, then serviced, then stroked on a permanent basis. The more you orient your business to their satisfaction the more contentment will be yours to enjoy.

The many faces of a customer

Your customers may be a diverse group, but they share an important similarity: They each have many different faces. The first time you see your customer's face, that customer looks remarkably unlike a customer and much more like a hesitant prospect, wondering if you're a rip-off artist and a charlatan, as so many other business people have proven themselves to be.

Your smile and direct eye contact, if you're in a business that allows you to utilize them, help alleviate the tension. Your learning of the prospect's name and telling the prospect your name form the first fragile threads of what can be a lifelong bond. Your comments about nonbusiness matters remove more of the natural salesperson-prospect pressure. You begin to relate as humans. That makes the prospect feel even better, and starts him or her on the wondrous transformation process from prospect to customer.

The wondrous transformation

You then ask the prospect questions with the prospect's interest foremost. The questions should either be about the prospect, which is probably his or her favorite topic, or they should be questions that can be answered in the affirmative. Often a few yeses start the momentum that leads to the sale. This all indicates to the prospect's unconscious that you are, indeed, interested in the prospect, not just in selling something and making money. Eventually, the prospect undergoes the transformation to customer. It happens suddenly, but the process has been going on a long time — through your marketing, your attitude, and your ability to earn the faith of the prospect.

After that transformation takes place, in your customer you will see a new face, relaxed and confident in you, taking on the guise of a repeat customer, coming back for more of what delighted him or her the first time, primarily because your follow-up marketing motivated the customer to return and to buy again.

It's a friendly face, that of the repeat customer. It is one that guerrillas bend over backward to see again and again. In time, with enough patient follow-up, that customer's face reminds one of a trusted and loyal sales representative for your company, extolling the virtues of your company to all who will listen. And many people will listen to positive word-of-mouth marketing. What is really happening is that your customer is but one cog in your ever enlarging word-of-mouth machine, a juggernaut that flattens your nonguerrilla competition. The enthusiasm the customer exhibits when talking of the benefits that result from patronizing your business is contagious — and always fascinating to hear, "easy listening," as they say in radio marketing circles.

Your word-of-mouth machine

Admittedly, more people will crane their necks to listen in on bad word-of-mouth marketing, but guerrillas are aware of the wildfire speed at which bad word of mouth spreads, so they take careful measures to avoid it. One of those measures is a customer-oriented recourse policy. That's the policy that keeps the promise of a long-term relationship alive even when the customer wears a face that scowls with displeasure. It dictates the impeccable service you will offer to rectify matters if things go wrong; it is deeply embedded in the mind of each of your employees; it is what your customers expect of you, although their expectations may have been dashed by less customer-oriented operations. After all, things do go wrong, even to guerrillas and their customers. But guerrillas with clear customer recourse policies change disgruntled customers into satisfied and repeat customers who will gladly tell the story of the spectacular service they received from you after they registered a complaint.

The anatomy of a customer

Customers come in all sizes, colors, creeds, genders, and ages. Although parents are influenced in purchase decisions by their children, the other end of the age spectrum, those over fifty, is where much of the money of the nation is controlled. Americans between the ages of fifty and seventy now control 77 percent of all financial assets — and that percentage is growing as more Americans pass the half century mark. The Census Bureau reports that households headed by people over fifty have $950 billion in annual income and have a net worth of $7 trillion. Of more interest to you might be the fact that these people control over fifty percent of discretionary spending. As you might expect, they care more for *quality and service* than price. But then, most people do.

Strangely, most companies continue to market primarily to people in the 18-to-34 age group, even though there are almost as many people in the 35-to-59 age group and even though that's the group with the disposable income. As reported in the *New York Times*, this occurs because most companies harbor the belief, proven wrong by research, that older consumers do not switch brands. The truth is that those over fifty have had at least thirty years of shopping experience; they know their way around and can't be wooed by anything but solid information. These people are very comfortable when they see things in writing. As a result, packaging is expected to become more informative. Service will become more obliging at the same time, and guarantees will be more substantive to attract older, more skeptical, and more practiced buyers. Already, the travel, health care, insurance, banking, and financial services industries are actively marketing to those fifty and older. You should consider doing the same if possible. Did you know that *Modern Maturity* — the second largest magazine in the United

States (*TV Guide* is first, but not for long) and one that goes only to people over fifty — is sent to 20 million people? Each week, another thirty thousand receive a subscription along with their membership in the American Association of Retired Persons (AARP). A marketing survey conducted in April of 1988 by Donnelley Marketing revealed that marketing professionals simply do not understand older consumers. They misunderstand their spending, their orientation to quality, and their willingness to change brands. Guerrillas understand these things clearly.

Regardless of the demographics of your own customers, an examination of them would show the following:

- They have specific values, probably different for each cus- **Truths** tomer. And they want you to be responsive to these values. **about your** They know that to do this, you must listen to them, then **customers** understand, analyze, and be sensitive to their needs.
- They don't mind paying a lot of money for a product or service, just as long as the quality merits the price.
- They were wary when they first purchased something from you. They've heard their share of horror stories of money wasted by bright people making purchases from shady businesses. They don't want to make a mistake. If they didn't, they feel an unconscious sense of gratitude to you. They respect you.
- They have expectations based on what you have communicated through your marketing, blended with their past experience as customers making other types of purchases.
- They have a basic human need for security, dignity, self-respect, and the respect of others. They are looking for a relationship of trust. The key to providing it — pay close attention here — is the fact that you and your staff *want* to build this trust. Please underline the word "want" in your mind. If your employees aren't capable of wanting a warm customer relationship, perhaps they should be in the employ of someone else.
- They have emotions and intellects. They all have experience as customers, as buyers. They may also have experience as

sellers. They want you to be customer oriented, though they've probably never used or heard that phrase. They have, however, dealt with businesses that were or weren't customer oriented. They're aware of the difference.

- They are not dependent on you. But you are dependent on them.
- They are doing you a favor by giving you the opportunity to serve them. You are not doing them a favor by serving. They are not interruptions of your work. They are the purpose of your work.
- They are not outsiders to your business. Instead, they're a vital part of it. They keep it alive. The more you realize that, the healthier they'll keep your business and your bottom line.
- They are not people to argue with. Do you think your business could win an argument with a customer? Guerrillas know that's like winning a skirmish but losing a war.
- They bring you their wants and needs. It is your job to fill and solve them. The better you do that job, the more profitable your business will be.
- They are the basis for most successful businesses in America — by their repeat business and their positive recommendations. Their withholding of repeat business and their negative recommendations are the basis of many business failures.

How does a business become customer oriented?

It begins with a desire to be that way and the energy to assure that the desire spreads throughout your organization and *remains there permanently*. Then it requires that everyone who deals with your customers remember the Golden Rule for Guerrillas:

A guerrilla's Golden Rule

ALWAYS TRY TO THINK LIKE YOUR CUSTOMER.

Walk a couple of miles in your customers' shoes. Look closely at your business through their eyes. Check out a business that is similar to yours. Honestly appraise to yourself its method of greeting you, its service, its identity. See what catches your attention. Determine what appeals to you. Focus on what would sell you. Why would it?

Although the guerrilla marketing attack is no game, you can play a game to hone your attack. Play the part of a customer. Have your employees, especially salespeople, play the role of customers. Encourage them to raise the objections and ask the questions that come from customers. If you fail to do this, you will forever lack an important perspective. When selling, listen carefully to those customers who challenge you, give you the toughest time. Show them great care. Address yourself to their needs, then win them over by your intense interest in what they want and your ability to provide it.

Listen when challenged

Empathy will be a valuable trait when you are faced with a customer complaint. If a customer takes the time to complain, that customer believes that you can do something about it. Be sure you reward this trust. Demonstrate as clearly as possible the specific steps you will take *personally* to make that customer completely satisfied.

Everyone knows that guerrillas must do their share of spying. Do some of yours by anonymously contacting your business with a request for information. Contact two competitors with the same request. Compare the attitude, the speed, the warmth, and the efficiency of the responses. Keep an eagle eye for follow-up. Fix anything that needs fixing. Outservice your competitors by every yardstick.

Spy on yourself

Reward and encourage innovation among your employees, especially in the area of customer service. One of my clients, a retailer, hands a crisp fifty-dollar bill to any of his employees who performs an act of outstanding customer service, especially an act of service that can be performed routinely, such as taking the time to explain extra features. It is important to show your gratitude for effective customer service concepts generated by your staff.

There is no question that employee relations mirror customer relations. Therefore, as a guerrilla, you've got to excel in employee relations. Treat them warmly and the treatment

Employee relations count

will spread. Recognize their specialness and they will do the same for your customers. Have high expectations and privately and briefly inform those who do not meet them how to reach the high standards for which you must become known.

This idea must be understood first by you, then by your management team, then by your employees. Your high expectations — and your company — are likely to be judged by the lowest standards attained. People do not recognize, remember, and talk about proper service and attentive attitudes as readily as they recognize, remember, and talk about poor service or uncaring attitudes. That's why customer love should be present throughout your company.

Plant the idea that "each one of us *is* the company," in the minds of all who work for you. And realize that customer satisfaction is not a cost, but an investment. For example, a local auto dealer provides shuttle service to public transportation for its customers. This takes time and money. But it results in repeat business and attracts new business.

Companies that enjoy remarkably high profits as a result of customer service have certain characteristics in common: They set astonishingly high standards of performance. They are obsessed with knowing what the customer wants. They know customer expectations must be understood and managed before they can be met and exceeded. They design their products and services to maximize customer satisfaction. They knock themselves out trying to be an easy company to do business with. They know the money they invest in customer service will pay off in satisfaction for the customers, profits for them. They repeat and repeat again that customer service is the responsibility of everyone in the organization.

Are the customers of today different from those of yesterday?

Of course they are. Customers are always changing. They are more informed, more sophisticated, more used to exceptional service from the growing ranks of guerrilla marketers.

The customers of today are also activists. If they're unhappy with your product or service, they're likely to tell the world, or at least their friends, relatives, co-workers, the Better Business Bureau, the local department of consumer affairs, the community consumer watchdog group, and the local newspaper. They may even contact *60 Minutes*. Their activism will result in better quality and better service throughout every industry. It forces shoddy companies to disappear or to clean up their act.

Today's customers are aware of details. They pick up on things like cleanliness, courtesy, convenience, packaging, and the care paid to the physical manifestations of your business. At Disneyland, one cannot help noticing the friendly Disney service people constantly walking around the grounds picking up wrappers, cigarette butts, bits of paper, anything. You *know* this is a well-run operation.

Keep in mind that you never have a second chance to make a good first impression. One of the best things about today's customers is their appreciation of top-notch service. They see it when it is there, and they tend to patronize businesses that consistently offer it. Once a customer sees that service is directed to his or her needs, wants, and values, that customer becomes part of a core group that will provide the foundation for your success.

Guerrillas who have the mind-set for a winning attack on the marketing front do not see customer service as an act or a deed. They see it as a continuing mode of behavior that leads to a long-term relationship. The relationship is a two-way street, with service traveling from you, and repeat business and a flow of referrals traveling from your customer.

The guerrilla's primary product

As the customer is the most important person in a guerrilla marketing attack, customer service and customer follow-up is the primary product. Customers don't buy products and services. They buy expectations.

Reward their expectations

Your ability to live up to, then beyond, these expectations is the real reason customers will return to you. In their conscious mind — and yours — your product or service is your primary product. But in their unconscious mind — and ideally in yours — the primary product is how well you reward a customer's confidence in you and how consistently you keep at it.

Ask yourself questions

To become skilled in the rendering of your primary product, a bit of introspection is useful. Ask yourself: How do I feel about being ignored? Of course, you hate it. So do *anything* to prevent your customers from being ignored — on the phone, in the office, in the store. Never let it happen.

Ask yourself: Do I like to be sold? Of course you do, even though you may not know it. We all love to be sold. We may dislike being pressured, but being sold is a different matter altogether. Be sure you are offering information, educating about benefits, showing the customer how you can improve his or her life — in short, selling.

Ask yourself: What companies do I deal with that make me feel the best? Have I patronized these businesses more than once? Have I recommended them to others? You can learn from the guerrillas around you. And remember that a company need not be new or small to qualify as a guerrilla.

Ask yourself: What is easier for my company to make, a first-time sale or a repeat sale? The answer is obvious.

And finally, never stop asking yourself: What can we do to improve customer service and to communicate with our customers?

Continually come up with answers and you're on the track to success as a guerrilla marketer, for you are proving through your actions that you understand the value of the most important person in your guerrilla marketing attack.

11
Thirty-three Marketing Myths

THE PATH to entrepreneurial success is mined with booby traps disguised as words of wisdom. Once you are actively engaged in marketing, you will be made privy to a galaxy of marketing facts. Although many will actually be true at the time you hear them, many will not, and a lot more will be true at the time you learn them, but false later on. It seems to take just about as long for a marketing maxim to become engraved in marketing minds as it does for that maxim to become a patent untruth.

Certainly, many of the facts of marketing remain true and will always serve as guides to the wise. But this chapter is about those truisms that never were true in the first place, or those that were true once but aren't true now. They were truths; now they are myths. Be alert for those myths. They are expensive. **Myths are** They are the cause of dashed hopes, and of failed products and **expensive** services.

To save you the expense and heartache of living by tired marketing axioms, I urge you to realize that these myths are truly mythical. Although you will continue to hear them and read of them, don't let them mislead you. Hope instead that they mislead your competitors. True guerrillas know that these "facts" are fables.

The speed at which marketing changes is one reason why facts are transformed into fiction. Another reason is that only recently has psychology pointed out the fallacies and opportunities in marketing. Business owners who do not keep up with marketing fall behind. And falling behind means living by myths pretending to be truths. Tuck these myths away where you keep the collected works of Homer, Aesop, and Mother Goose. Treat them with the same respect: They may be enjoyable to read, but they're no way to run a railroad, let alone a marketing campaign.

I have mentioned some of these marketing truisms earlier. However, so intense is my distaste for them, and so common my experience with clients who based hopeless campaigns on them, that they bear repetition.

Does this mean that *all* of these are *always* to be taken as myths and never followed? No, it doesn't. There will be exceptions. There are supposed to be exceptions in marketing *some* of the time. Leo Burnett used to remind his crew: "It is important to know the rules so that you'll also know when to break them." To that I add: "It is important to know the myths so that you'll be aware of them when they appear before you, masquerading as the rules."

Myth #1: It is good to have a great deal of white space in advertisements, brochures, and other printed material.

Truth #1: Your customers and prospects are not interested in white space. They are interested in information, in what your product or service can do for them.

When you use white space where you could be putting information in the form of features, benefits, or specific ideas of interest to your prospects, you are wasting money. Art directors who use large amounts of white space to surround a message or visual element are relying on an unimaginative and expensive attention-getter rather than a powerful graphic image. **Where's the beef?** Printed materials should attract attention with substance rather than emptiness.

There is little question that a great expanse of open space adds to the aesthetic appeal of most printed materials. But there are far more intelligent ways of using expensive marketing space aesthetically other than leaving it blank.

Myth #2: People are likely to open thick envelopes.

Truth #2: In the halcyon days before the daily direct-mail clutter, thick envelopes did attract readership with their hidden promise of a storehouse of information and offers. But these days, those thick envelopes appear to be robbers of precious time, filled with hype and confusion.

However, you might consider intriguing the recipient of your mail by enclosing some interestingly shaped object in your envelope, such as a ballpoint pen with which to complete a response device. The intrigue would not seem to be an intruder, and the strange feel to the envelope would not feel like hype. Adding a special brief note, called a "buck slip," to your classic **Buck slips** mailing package doesn't hurt either. Buck slips usually are folded in half and say on the outside: "If you read nothing else, read this!" Then they open to reveal the main idea — the basic heart of the offer. Buck slips increase readership without adding bulk.

These days, bulk adds to cost and cuts down on readership. Avoid a thick envelope unless you've got a darned good reason for sending it and your outer envelope has a high-potency message to prevent the recipient from following a natural instinct to toss the intrusive thing away unopened.

Myth #3: Use short copy. People won't read long copy.

Truth #3: This is one of the most dangerous, costly, and silly myths in marketing. People will read long books, long articles, long letters. They'll read whatever interests them, and the more it interests them, the more of it they want.

When presented with a marketing offer, they want enough information to make an informed decision to buy — or not to buy. If they don't have enough information, they won't buy. Remember that in all marketing. Give people more information than they need, and they're likely to buy or not to buy. Withhold that information, and they're very likely not to buy.

Be not self-conscious   There's a self-consciousness about marketing in the minds of most nonguerrilla marketers. Somehow, they feel it beneath their dignity to market, to sell, to risk rejection. So they produce marketing materials that are sparse in copy, feeling in some crazy way that perhaps if they keep the copy short enough, people won't realize it is marketing. But people do know it is marketing even if it has no copy. Those who don't care what is being marketed won't care about the marketing. Those who do care what is being marketed will want to know as much as possible about it. The last thing you want is to deny them that information.

The most appalling application of this myth is in brochures with short copy. Brochures are the ideal medium for long copy. They are expected to furnish details. Prospects generally request them or pick them up simply because they want information. But alas, many brochures malnourish the minds of inquisitive prospects with a few pretty pictures and a couple of paragraphs of prose. Brochures must be treated as opportunities to sell, not as opportunities to be coy and secretive.

If you feel self-conscious about marketing, maybe it's a good idea to devote your time to something about which you feel more comfortable, and leave marketing to guerrillas who feel fortunate to be able to supply information to possible lifelong customers.

Myth #4: It is costly to purchase radio and television advertising time.

Truth #4: Radio and TV advertising time does not have to be costly, is coming down in price regularly, is delightfully negotiable, and often proves to be a superb value. This is especially true of television, now available in some markets for $3.00 per minute.

Myth #5: It is expensive to produce an effective TV commercial.

Truth #5: Frankly, it *is* expensive to produce an effective TV commercial. But it doesn't have to be. Although in 1987 the average TV production cost for a thirty-second spot was $93,000, and some were effective, many of those high-priced commercials were stinkers. During that same year, scads of thirty-second TV spots were produced for less than $1,000. Some of those low-priced commercials worked as poorly as the high-priced stinkers, but others were highly effective. The idea made the difference, not the cost of the production.

The idea makes the difference

Great television is based on an interesting idea made visually appealing and clear to the prospect, and abetted by interesting sounds, generally a blend of words, music, and sometimes sound effects. It does not take $93,000 to produce thirty seconds' worth of great television.

It does take a keen imagination, a feel for the product or service, a sense of what the prospects want, a desirable benefit for your audience, a believable presentation, and an ability to resist the temptation of unnecessary special effects, high-paid presenters, glamorous locations, and hucksterish gimmicks.

Many who are given the responsibility of producing TV commercials see their task as skimming the leading edge of TV technology and dazzling the viewers with entertainment. Someone should remind them that their task is to make the product or service interesting and desirable. Although there's a lot of show biz in marketing, there's a lot more, too. Solid selling ideas cost no more to communicate than shallow ideas. Often, an expensive commercial is a device to hide the lack of a real idea or consumer benefit.

Although you have my permission to spend $93,000 producing a TV spot, you have my encouragement to spend $92,000 less. Do it by producing more than one commercial at a time, holding a detailed preproduction meeting, making sure everyone involved knows what to do on the day of the shoot, and shooting without sound, adding it later.

Myth #6: Sell the sizzle and not the steak.

Truth #6: Sell the solution and not the sizzle. The easiest way to sell a product is to offer it as the solution to a problem. If you tend to look for the sizzle rather than the problem, you are looking in the wrong direction. Your prospects might appreciate the sizzle, but they'll write the check for the solution. Do all in your power to identify problems that your prospects have, then position your product or service as the best solution to that problem.

If you think solutions, you'll market solutions. If you think sizzle, you'll sell sizzle. These days, people love sizzle as much as ever. But given a choice of purchasing sizzle or a solution with their discretionary income, customers will put their money on the solution every time.

The guerrilla marketing attack calls for you to focus on the prospect, then see the problem. When you present your offering as its solution, you follow the path of least resistance to the sale. With little resistance, your attack succeeds.

Myth #7: *If you have the right offer, price, mailing package, and mailing list, your direct-mail effort will be successful.*

Truth #7: I wish it were that easy. But *timing* must be factored into your formula; it is always important. If your direct mailing is top quality but your timing is off, you'll miss the sale. One of your most valuable allies in a fierce marketing environment will be realism in direct marketing. Realism is a mountain of mail competing with yours, a mountain that arrives every day but Sunday. Ideally, your mailing will not be part of the mountain because your timing was right.

Realism in direct mail

Myth #8: *Great marketing works instantly.*

Truth #8: Great price-off sales work instantly. Great limited-time offers work instantly. But great marketing is not made up of price-off sales and repeated limited-time offers alone.

Great marketing is made up of creating a desire for your offering in the minds of qualified prospects. It often is peppered with price-off sales and limited-time offers. But when a greedy nonguerrilla embarks on a program of such fast-buck schemes, he or she quickly learns that the public makes a purchase, then waits for the next sale. If you don't have one, they don't buy.

Myth #9: *People have strong opinions about marketing.*

Truth #9: A study involving five thousand people throughout the United States, designed to determine their attitudes about marketing, revealed that the majority (67 percent) didn't give enough thought to marketing to have any opinions about it. I have opinions and you have opinions. But the public has more important things to think about than marketing.

Myth #10: Marketing should entertain and amuse.

Truth #10: Show business should entertain and amuse. Marketing should sell your offering. It seems to me that this particular myth is one of the most widespread in marketing.

A widespread myth

To see how entertaining and amusing marketing detracts from the basic appeal of the product or service, you have only to check your newspaper or favorite magazines for funny, punny, or rhyming headlines. Or watch TV for those commercials that you remember so well, but somehow forgot what product they advertised. The advertising community perpetuates the cult of the clever marketing message by means of awards shows where glitz and glitter, humor and originality, special effects and knock-'em-dead music win awards for marketing. Those awards should be given for profit increases, nothing else. Being one who is as guilty as you can get of winning marketing awards for the wrong reasons, I now publicly renounce all those awards that did not lead to glittering bottom lines for my clients.

Myth #11: It is not possible to be creative when marketing certain products and services.

Truth #11: When's the last time you saw the word "pshaw!" in print? You are seeing it now. It used to be believed that there was no room for creativity in insurance marketing, no place for creativity when marketing professional services, no way to put forth captivating bank advertising. This myth has been exploded sky high by companies who realized that creativity comes in influencing human behavior, not in winning prizes or grins. Today all marketers, from doctors to grocers, have the opportunity to be ultracreative. Once past the barrier of thinking that creative marketing relates to marketing as an art rather than a business, companies and individuals in all industries can be creative in their message, their media, their offer, their mailings, indeed, in all their marketing.

Myth #12: Nobody pays attention to TV commercials.

Truth #12: Well, I pay attention to TV commercials as do other marketing people. But more important, 39 percent of Americans said in a 1987 study that they actively watch all the commercials that appear in and around the shows they watch. I tell you this because earlier I warned you that people do not watch marketing; they watch only what interests them, and sometimes it is marketing. That is true. Also true, however, is an inborn fascination with the medium of television, which includes its commercials.

As you might expect, TV viewers represent a hodgepodge of attitudes and actions: zapping through the taped commercial, muting the commercial, checking other channels during the commercial, talking during the commercial, leaving the room during the commercial, and paying rapt attention during the commercial. Don't believe that nobody pays attention. Four of ten viewers do. And four of ten represents a lot of people, many of them prospects for an offering, or else why would so many businesses be using television in the first place?

Myth #13: Advertising on television gives your business a glamorous identity.

Truth #13: Even though TV is part of the world of show business, there is little glamour associated with advertising on

Study after study has shown absolutely no association between people remembering your marketing and buying your offering. The same studies have proved that it matters not whether or not folks like your marketing. All that matters is whether or not people are motivated by your marketing to purchase your offering. Procter & Gamble test commercials to see if people remember the main point of the commercial, but not necessarily the commercial itself. Many memorable commercials have been quoted back and forth by commercial aficionados and have won baskets of awards while sales of the product were sliding on a precariously downward slope. It's not how memorable you make the marketing; it's how desirable you make the product or service.

Myth #16: Public relations stories have a short life span.

Truth #16: If a magazine or newspaper runs a favorable story about your business, you can read it, grin widely, then let it fade from the public's memory — or you can do several things to extend its life span. Here are a few tacks open to you:

Extending the life span

1. Make reprints of the story and include them in your next mailing.
2. Make one large reprint of the story, frame it, and hang it in your store or reception area.
3. Include a reprint of the story as part of your next brochure.
4. Include excerpts from the story in your advertising campaign.
5. Mail the story, along with a press release, fact sheet, and your photo to a contact at a local radio or TV station. News feeds upon itself.

Many a guerrilla gains business today from a marvelous PR story that ran a dozen years ago. A public relations story has a short life span only if you let it die. Guerrillas are masters of publicity resuscitation.

Myth #17: Bad publicity is better than no publicity at all.

Truth #17: Bad publicity is bad for your business. No publicity is a lot healthier for you. People love to gossip, especially about businesses that have done something so bad that it got written up in the paper or exposed on the TV news. That's why bad word of mouth spreads so rapidly.

Perhaps for a no-name politician seeking any kind of publicity, bad publicity is better than none — simply for the sake of name recognition. But I'm not too sure about that. I am sure, though, that bad publicity is something that gives joy to no self-respecting guerrilla.

Myth #18: Small businesses cannot afford to advertise in national magazines or large metropolitan newspapers.

Truth #18: Today, small businesses can easily afford the regional editions of many national magazines as well as the zone editions of many large metropolitan newspapers. The world of media is becoming more and more oriented to small businesses, especially those run by guerrilla marketers.

Myth #19: Word-of-mouth marketing is all a great business needs.

Truth #19: Amazingly, some otherwise well-informed business people believe this myth to be true. Here and now, I implore you to understand that it is hardly ever true.

How will the great business get people to spread the word

in the first place? Marketing is the answer. How will people hear of the small business when it is new? Marketing is how. Where will the people come from — those who will make all the referrals? They will come from compelling marketing.

It is true that great marketing can attract so many people to a great business that word-of-mouth marketing is active and effective. But that takes time. It takes coddling of customers, customers who came in because of marketing. And anyhow, that customer-coddling is marketing.

I have had a few clients who were able to discontinue their marketing because they reached the limits of their growth. But I have witnessed others who thought they could discontinue marketing only to find that a competitor took their customers away from them.

The guerrilla marketing attack demands that you offer so much quality and service that word-of-mouth marketing becomes one of your most devastating weapons. It should always be part of your arsenal, and you should do all in your power to encourage and promote it. But I do not recommend that you rely on it solely. The bankruptcy courts are littered with businesses that felt they could save on marketing by leaving everything to word-of-mouth marketing. Life just doesn't work that way. And guerrillas know it.

Word of mouth can't do it all

Myth #20: *The purpose of marketing is to generate maximum sales volume.*

Truth #20: The purpose of marketing is to generate maximum profits. A high and increased sales volume is nice, but not at the expense of profits. I know of a business that decided it could sell a whale of a lot of furniture if it gave away free goods to purchasers. A massive ad campaign was launched and prospects came to the showroom in droves, buying everything in sight and taking the store up on its offer of valuable freebies.

When the dust settled, the store owner was jubilant. He had hit an all-time high in sales. He had moved more merchandise faster than ever before. It wasn't until a few weeks later that the controller of the company, the guy in charge of the purse strings, sadly announced that the company lost money during the promotion. The cost of the marketing and free goods was disastrously high. The company owner quickly became a marketing guerrilla. After he licked his wounds, he repeated the promotion, this time offering higher priced items for sale and giving away lower priced freebies. This time, he made money.

Almost any business owner can raise sales volume with enough marketing and a tempting enough offer. But it takes a guerrilla to add an upswing to the profit curve through marketing. Increased sales are nice, but increased profits are why guerrillas love marketing.

Myth #21: Quality is the main determinant in influencing sales.

Truth #21: Quality is the second most important determinant in influencing sales. Confidence in the business is the main determinant.

Nobody wants to buy the best product if it comes along with the poorest service. People aren't interested in quality if they have to sacrifice self-esteem. Just as word-of-mouth marketing is an integral part of the guerrilla marketing attack — but not the only part — quality products and services are also key elements in the attack — but not the only elements. They must be present.

Customer service must also be present. A friendly attitude must be displayed. The customer must be singled out as special. That customer should be provided with a selection, with convenience, with flexibility in paying for the purchase, with the feeling of a good value. Prospects become customers of businesses that offer credibility — in decor, attire, displays, marketing, employees, and especially in their reputation for offering value. Those items *plus* quality influence sales. Unfortunately, quality alone won't do the job.

Myth #22: *It is important always to have a sale of some kind.*

Truth #22: Reduced-price sales can be very dangerous. They result in a quick rush of increased volume, but that quick rush becomes addictive, causing the business owner to have more and more sales. This is when sales become counterproductive.

Just say no

Prospects wait for a new sale before making their purchase. If the business owner waits to have another, the prospect will take his business elsewhere. Sales also appeal to a class of people who tend to shy away from the most profitable merchandise and instead concentrate on the cheapest. From such customers come unhealthy customer bases. People who patronize a business because of a sale are not likely to be repeat customers. Instead, they are a fickle lot who go wherever the prices are lowest.

Price-off sales eat into profits, often undermine credibility, and serve as excuses for the lack of a creative selling idea. If you can build a business without ever having a sale, your customers will be more loyal, will patronize your business more frequently, and will spread word-of-mouth marketing referrals far beyond "low price."

I don't recommend steering completely clear of price-off sales. They are a valuable weapon to a guerrilla. I only recommend avoiding dependency upon these kinds of sales.

Myth #23: *It makes a lot of sense for a small business to retain the services of an advertising agency.*

Truth #23: It makes little sense for a small business to retain the services of an ad agency. Better work at more reasonable prices is available from a marketing consultant.

Naturally, an advertising agency tends to view advertising as the most important cog in the marketing cycle. Frequently, the exact opposite is true. The small business needs a direct-mail campaign tied in with a telemarketing effort. Most ad agencies are not set up to establish such a program. Many have never written a direct-mail package or even seen a telemarketing script. After all, they are *advertising* agencies, and direct mail plus telemarketing doesn't count as advertising.

Although the key people in an advertising agency may make a presentation to a small company in order to secure the business, all too often assignments for the company are given to the junior members of the ad agency. Being a business, the agency is aiming for maximum profits, and it just isn't profitable to put their highest-priced executives on the business for a fledgling company. This is not true of all agencies. And it also doesn't mean that junior ad people are less talented than their more experienced elders. But it is true most of the time in most of the agencies.

Advertising agencies maintain high overheads, usually to service their larger clients. Small businesses compensate the agencies and pay for some of the overhead from which they derive zero benefits. If the agency has a modern research facility because some *Fortune* 500 client requires it, your small business may help finance the facility even though you do not utilize its services. Your small business also helps pay the tab for the high rent that many ad agencies must pay in order to attract blue-chip clients. Check out your local agencies to find exceptions to these heinous charges. At the same time, check out marketing consultants who place advertising in its proper perspective, who charge only for what they deliver, and who will personally work on your marketing for you, utilizing free-lance talent when necessary.

Please understand that I think very highly of advertising agencies. I really do. It's only that I think even more highly of you.

Myth #24: You can save money by producing your marketing materials within your own company, using your own people.

Truth #24: The way to view your production of marketing materials is to recognize it as a place to protect your marketing investment. The brightest marketing strategy can be completely undone by any hint of unprofessionalism in your brochures, ads, signs, letters, and so forth. The whole idea is to gain confidence, and you won't gain any with home-grown creative materials — unless they are superb.

With all your marketing comes a metamessage — a non-verbal statement about your entire company. That metamessage is carried in the feel and tone of your words and pictures, your type style and paper stock, your ad size and media selection. It is every bit as important as your stated message. Small businesses that convey less than excellence in their finished materials are exhibiting the ultimate in unguerrilla behavior: They are saving small sums of money while sabotaging large sums.

Your metamessage

A few businesses successfully conduct their entire marketing operation in-house: strategic planning, writing, graphics, media buying, telemarketing, and all production. I give you my solemn promise that these are the exceptions to the myth.

Myth #25: Once your business has a solid customer base, it can cease marketing.

Truth #25: You can cut down drastically on your marketing to the universe, that entire area or industry that comprises the first of your three markets. You might even reduce the marketing to your second market — your prospects. But you should always market to your customers, your third and most poten-

Why customers will desert you

tially lucrative market. Even if your offerings completely satisfy your customers, some of these beloved citizens will still desert you. It won't be your fault. It will be because they have moved away, died, been wooed away by a competitor, lost their purchasing power, or one of many other reasons that do not reflect adversely on you.

You've got to treat marketing as a continuing need to take up the natural slack. You really and truly can cut down on your marketing expenses once your customer list is large enough. By investing just enough in prospects to replace customers lost to fate, and investing generously in marketing to your own customers, you can decrease your marketing investment while your profits soar.

Myth #26: If a prospect says he or she wants some time to "think it over" before making a purchase, they actually will think about it, and they'll probably come back to buy.

Truth #26: They'll buy all right, but not from you. When a person selling something deals with a person who is shopping, a sale will be made. Either the prospect will buy the product or service, or the seller will buy the prospect's story about why that person can't buy right now. You've got to figure that once a person comes into your place of business, they do want to buy. And if they leave, they don't want to buy from you, regardless of how much they'll think it over. Yes, there are a few exceptions, but only a few. The time to close a sale is during or immediately after the sales presentation.

Myth #27: In order to market effectively, you'll have to spend more than you can afford.

Truth #27: The guerrilla marketing attack can be tailored to the reality of any budget. You can market effectively with any size budget. The degree of your effectiveness can be affected by your willingness to see your marketing funds as an investment rather than an expenditure.

If you are not experienced in business, you may not realize the vast importance of marketing to your enterprise. Many intelligent people fail when they launch their own business because they never learned the importance of marketing. True entrepreneurs are willing to take the risk inherent in the concept of marketing. They view marketing as something they cannot afford *not to do.* You may have to invest more than you planned and more than you want to, but if you invest it in a guerrilla marketing attack, it will not be more than you can afford.

Myth #28: You should use as many of the media as possible.

Truth #28: You should use as many of the media as you can use effectively. If you can't use a medium properly, stay away from it. "Properly" means produced in a manner to inspire confidence, and run with enough frequency to capture attention and make an impact. Properly means running ads that do not portray you as less than your competition, either in your message or metamessage.

It is better to use two media properly than to use four in a way that spreads you too thin. Guerrillas know that there will be synergy between all the media they employ, so they employ as many as possible, but they know where to draw the line.

Myth #29: Repetition of a marketing message is boring.

Truth #29: It may be boring to you, but it won't be boring to your prospects and customers. Repetition implants your benefits in the unconscious minds of your prospects, and reaffirms those benefits in the conscious minds of your customers. Repetition does not bore these wonderful people.

Myth #30: Marketing is too complex for you to control.

Truth #30: Guerrilla marketing is not too complex for you to control. Guerrilla marketing removes the hocus-pocus from real-life marketing and reveals it for what it really is — replete with opportunities to serve customers and improve their lives. Each of the opportunities requires time, effort, or imagination, but they are understandable and you can take advantage of them. Marketing activities, such as sign posting, stamp affixing, or letter writing, that you cannot or do not want to do, you can delegate. You can locate a pro with a track record of success and let him or her excel at marketing for you while you excel at your primary business. Because you are operating from a simple strategy and calendar, utilizing a simple system of evaluating, then selecting weapons, you are in complete control of marketing. Never again will you shy away from aggressive marketing simply because you feel that you have no control over it.

Myth #31: *Large businesses cannot use guerrilla marketing.*

Truth #31: As more and more *Fortune* 500 companies are discovering, large companies can and should use guerrilla marketing at every turn.

The principles of commitment, investment, consistent, confident, patient, assortment, and subsequent are as applicable to the biggies as the babies. Huge corporations are learning the power of the mini-media. High-placed executives are using personal letters to influence multimillion-dollar sales. Telephone calls to key people are replacing older and more standard marketing approaches.

Large companies are able to evaluate the list of 100 weapons and find more to use than they realized were available. With such high stakes as survival and profits, it makes sense for large companies to capitalize on every opportunity and employ tactics not found in textbooks. The idea is to market like a guerrilla, regardless of the size of your company.

Myth #32: *Offering free gifts is beneath your dignity and that of your prospects and customers.*

Truth #32: There is no loss of dignity involved in giving a free gift that will be perceived as valuable by your prospects and customers. You've got to use good judgment when selecting the gifts. For instance, if a medical clinic wanted to market by offering a free gift to patients or people who came in for specific medical tests, it would be highly unprofessional for them to give fancy books of matches or ashtrays featuring the name of the clinic. But it would be in keeping with the dignity and

Right gifts and wrong gifts

standards of the medical profession if they gave handsome free books about first aid treatment.

Banks, insurance companies, and corporations of all types and sizes can find items that would make splendid gifts for their prospects and customers. Many of these gifts would actually add to their credibility rather than detract from it.

Myth #33: All that really counts is moving the goods and earning an honest profit.

Truth #33: Good taste and sensitivity also count. Marketing, as part of mass communications, and as an integral component of the information age, is part of the evolutionary process. It educates, informs, announces, enlightens, and influences human behavior. Because it does, it has an obligation to offend no one, to present its material with a sense of taste and decency, to be honest, and to actually benefit customers. Guerrillas are out there to outmarket the foxiest competition, but to win battles fairly waged without affronting anybody.

The obligation of marketing

By exposing these common myths, I have tried to protect you from the fables that would stand as barriers between you and success. With marketing growing with such rapid strides, the body of marketing myths is keeping pace.

Along with these 33 myths, be on the lookout for others that go against your common sense, that rub your instincts the wrong way. Guerrillas are always on the alert for mindless traditions in marketing.

12
Maintaining the Attack —
and the Profits

THERE ARE THREE PHASES to the guerrilla marketing attack. To win consistently healthy profits, you must succeed at all three. Two out of three isn't enough.

1. The first phase is the *planning*.
2. The second phase is the *launching*.
3. The third phase is the *maintaining*.

You now have the insights and know the tactics to get your adrenaline flowing and to *plan* an ambitious, aggressive, and multifaceted attack. The momentum you sense when you hold in your hands a finished guerrilla marketing plan and calendar will carry you forth with the vigor to *launch* it successfully.

As much time, energy, imagination, and money as this took, it was the easiest part of the attack. The hardest part is yet to come. The hardest part is to *maintain* the attack constantly. This is the key to the attack, to the concept of guerrilla marketing, to the presence and size of your profits. Because marketing is a living process, rather than a single lifeless moment, it requires attention, nurturing, time.

The hardest part

Your goal must be to set your marketing into motion. If you're a guerrilla, you'll do it in a way that assures the maintenance of that motion. Develop your plan and calendar. Create and produce your marketing materials. Order your advertising time or space, and send the proper materials, along with instructions, to the media. Plan your direct-marketing program, select your lists, and produce your packages, scripts, and other weaponry. That's the first part of maintaining your attack.

Every single week The second part is to set aside a certain amount of time each week for sales training, feedback from employees who deal with the public, and attending to the other details of marketing. The little time this takes each week is really all you must put into the process unless you are augmenting your marketing attack with new information for your prospects and customers.

If you — or a person you've made the designated guerrilla — don't devote this weekly time to marketing, the proper maintenance of a full-scale attack will be almost impossible. You'll think you've attended to all the details, but some will still be less than 100 percent. Improve them.

On the other hand, if you spend more than a few hours a week once the attack has been launched, you're probably doing something wrong. Unless you are creating new materials, making major changes, or preparing for a seasonal marketing effort, you won't really have to devote too much of your time to marketing. Great marketing deserves full-time attention, but is not a full-time job.

Maintaining a guerrilla marketing attack destined for long-term victory requires ten things.

1. A PERSON

Maybe you will be the person to plan, launch, and maintain your company's attack. If so, I hope you relish the opportunity. But if not, be certain that some other person in your business takes on and looks forward to the attack. See to it that the person you select has the guerrilla personality traits of patience, aggressiveness, imagination, sensitivity, and ego strength.

2. A COMMITMENT

The person directing the marketing attack and the person who owns the company must both be sincerely committed to planning, launching, and maintaining the attack — *especially maintaining the attack*. If both of these people are the same person, that's fine. If not, mutual commitment to victory is essential. Without it, marketing will become for you the futile and expensive exercise in mindless spending that it is for so many of America's businesses — both small and large.

3. A TIME

Schedule time to do all the mental work and time to handle the physical work of launching a guerrilla marketing attack. While you're at it, also schedule a regular time each week when you can devote new energy and imagination to maintaining your guerrilla marketing attack. This regular weekly marketing time will reward you far in excess of your hourly income. The marketing time can be a period in which you alone consider and augment your marketing activities — or it can be a marketing meeting involving salespeople, production people, telephone answering people, and all other employees who have any contact with the public. You must tailor your marketing time to your current situation and objectives. But don't spend too much time at it once your attack is launched.

4. AN ANALYSIS

Analyze all the 100 weapons in the guerrilla marketing arsenal. Although you will do this prior to writing your marketing strategy, you should also do it at least twice a year — to sharpen your analysis and restock your arsenal as necessary. The guerrilla marketing attack begins with this analysis, and is fueled by a desire to utilize as many weapons as possible. Unless you can use them the right way, put them off until you can. For example, if you can't afford a brochure and telemarketing follow-up yet, put off your plan for a direct mailing. The idea is not merely to use all your weapons, but to use them so that they produce the results you desire: increased profits. You want them to be effective for your company. Direct marketing, however grand an opportunity, is a drain on your time and your budget if you do it less than superbly. TV is a powerful medium, but not if you have a poorly produced commercial. Guerrilla marketing tools are not toys; they are lethal weapons and must be used right or not at all.

5. A STRATEGY

Based on a clear-headed appraisal of your potential marketing arsenal, your industry, your competition, your budget, your goals, your market, and yourself, create a seven-sentence strategy that can guide your efforts for the next five to twenty years. Be certain that everyone in your company who deals with any member of the public reads and understands this strategy. You'll find that many who understand it can show you ways to strengthen it. You'll also find that some of your competitors **Expect to be** will do all they can to copy your strategy. Worry not. Many **copied** of your prospects will recognize the copycatting and patronize you as a result. They'll figure if you're worth copying, maybe you're onto something good.

6. A CALENDAR

Your 52-week guerrilla marketing calendar will undergo a *remarkable increase in value each year* as you pare the marketing efforts that didn't achieve impressive results while increasing those efforts that did. This provides increasing savvy and muscle for maintaining and continuing your attack. That's why many companies make a new 52-week calendar every 26 weeks. After three years, many guerrillas consider this calendar to be one of their most precious assets for long-range planning. They **More valuable** find they can do something about those one-time bad months. **each year** Each year, their calendar gains precision as they gain battle experience.

7. THE TRACKING

Where did each of your customers learn about you? Where did your prospects who are not yet customers learn about you? Unless you conscientiously track the sources of all new people who contact your business, you won't know the answer to these very important questions. Yes, I know that tracking is interruptive, but guerrilla marketing requires honest effort. Each call to your business, visit to your store, coupon sent to your company, response to your direct marketing represents a wasted marketing effort unless that contact is tracked. I'll grant you that tracking is not the most enjoyable part of marketing, but it is one of the most necessary parts.

Not fun, but necessary

8. THE TRENDS

Although your competitors will probably react to you, rather than the other way around, you've got to be ready to react to new trends in marketing, in your industry, in business. That means you've got to keep up with marketing, your industry, and life in general. New trends are opportunities for you to add technology to your arsenal, psychology to your message, energy to your attack. Capitalizing on new trends is the tactic you'll use to keep the competition guessing, to prevent your own attack from being predictable. Books are wonderful for providing you with insights. But magazines and newsletters furnish the front-line intelligence from which action springs. Keep up or you'll lose out.

9. THE CUSTOMER

Consistent contact with the customer subsequent to the sale is what will create the most profits for your company. Maintenance of that contact will earn a fortune for you. Failure to maintain contact will lose a fortune. Based on your own industry and customers, select the most appropriate methods of maintaining customer contact, maximizing repeat business, and gaining the most referral business. This will be relatively simple if you *think in terms of the customer.* What does the customer want? Need? Dream of? Fear? Find ways to serve your customer and regularly offer new forms of assistance in the form of services, products, and information. There will be a direct relationship between the quality and consistency of your customer contact and the size of your bankroll at the time you sell your business or retire.

10. A MOTTO: DO IT NOW

An old Chinese proverb suggests that the best time to plant a tree was twenty years ago. The second best time is today. Reading about the guerrilla marketing attack may be fun, but it won't hurt your competitors, increase your share of mind and market, or energize your cash flow until you plan, launch, and maintain it. If I ran a company that was doing less than 100 percent on the marketing front and I encountered information **Spring into** such as this, I'd spring into action — either in the form of **action** devoting office time to handling the planning paperwork or in the form of calling a meeting and appointing the person or people responsible for the care and feeding of a fierce attack on all marketing fronts. It's not the kind of thing you mull over. It's something you activate. And you do it *now* if you're a guerrilla and not a bystander.

More small businesses are being started now than at any other time in history. The success rate is rising as they are learning more and more about marketing. Consider:

• In 1970, 7 percent of all working Americans were self-employed. In 1986, that figure rose to 11 percent. By 1990, it will be 15 percent. According to the Department of Commerce, the self-employed work force is growing nearly four times faster than the salaried work force. Many of these former jobholders are your current or future competitors. Maintaining your attack will help keep them at bay.

- In 1950, we were creating new businesses at the rate of 93,000 yearly. In 1980, the rate rose to 600,000 yearly. By 1990, it will be nearly 1,000,000. Keep up your attack.
- Today, there are 11 million businesses in the United States. Of those, 10.8 million are small businesses. All the more reason to ceaselessly be on the attack.
- Some small businesses obtain capital from venture capitalists. In 1977, the venture capital pool was $39 million. Three years later, it increased to $900 million. At this moment it is well past the $1 billion mark and still enlarging rapidly. In his book *Megatrends*, John Naisbitt terms the phenomenon "an entrepreneurial explosion."

Surviving in the 1990s

In the fierce business climate of the 1990s, only the most marketing-minded will survive. Those will be the guerrillas, the ones who have made their company name a brand name in which their prospects have confidence. They will find it necessary to market with maximum energy throughout the life of their business. But they'll enjoy it with the same gusto as a gold-medal Olympic athlete about to enter an event.

Guerrillas know how to win. They've won before. They'll continue winning. They know the work is ceaseless and often hard. But the rewards are lush and nourishing. And they know that if they don't win, they'll lose.

The wisest thing that you can do to assure your membership in the ranks of the victorious is to launch your attack as soon as possible. I am talking about your marketing, your business, and your future. If you don't embark on an all-out guerrilla marketing attack right now, there's a very good chance that you'll begin to fall behind right now.

Information Arsenal for Guerrillas

Index

Information Arsenal for Guerrillas

Applegath, John. *Working Free: Practical Alternatives to the 9-to-5* **Books**
  Job. New York: AMACOM, 1981.
Baty, Gordon. *Entrepreneurship: Playing to Win*. Reston, Va.: Reston Publishing, 1974.
Bayan, Richard. *Words That Sell: A Thesaurus to Help Promote Your Products and Ideas*. Chicago: Contemporary Books, 1987.
Bencin, Richard L. *Strategic Telemarketing*. Philadelphia: Swansea Press, 1987.
Bencin, Richard L. *The Marketing Revolution*. Philadelphia: Swansea Press, 1985.
Benn, Alec. *The 27 Most Common Mistakes in Advertising*. New York: AMACOM, 1978.
Benson, Richard V. *Secrets of Successful Direct Mail*. Savannah, Ga.: Benson Organization, 1987.
Bly, Robert W. *The Copywriter's Handbook*. New York: Dodd, Mead, 1985.
Bobrow, Edwin E., and Dennis W. Shafer. *Pioneering New Products: A Market Survival*. Homewood, Ill.: Dow Jones–Irwin, 1986.
Bonoma, Thomas V. *The Marketing Edge: Making Strategies Work*. New York: Free Press, 1985.
Bove, Tony, Cheryl Rhodes, and Wes Thomas. *The Art of Desktop Publishing: Using Personal Computers to Publish It Yourself*. New York: Bantam, 1986.
Breen, George, and A. B. Blankenship. *Do-It-Yourself Marketing Research*, Second Edition. New York: McGraw-Hill, 1982.
Buell, Victor P. *Handbook of Modern Marketing*, Second Edition. New York: McGraw-Hill, 1986.

Daniells, Lorna M. *Business Information Sources.* Berkeley: University of California Press, 1976.

Day, William H. *Maximizing Small Business Profits.* Englewood Cliffs, N.J.: Prentice-Hall, 1978.

Dean, Sandra Linville. *How to Advertise: A Handbook for Small Businesses.* Wilmington, Del.: Enterprise Publishing, 1980.

Dible, Donald M. *Up Your Own Organization.* Santa Clara, Cal.: Entrepreneur Press, 1974.

Ehrenkranz, Lois Beekman, and Gilbert R. Kahn. *Public Relations/ Publicity: A Key Link in Communications.* New York: Fairchild Books, 1983.

Eichenbaum, Ken. *How to Create Small-Space Newspaper Advertising That Works.* Milwaukee: Unicom Publishing Group, 1987.

Eicoff, Al. *Eicoff on Broadcast Direct Marketing.* Lincolnwood, Ill.: National Textbook Company, 1987.

Feinman, J. P., R. D. Blashek, and R. J. McCabe. *Sweepstakes, Prize Promotions, Games and Contests.* Homewood, Ill.: Dow Jones–Irwin, 1986.

Fisher, Peg. *Successful Telemarketing: A Step-by-Step Guide for Increased Sales at Lower Cost.* Chicago: Dartnell, 1985.

Foote, Cameron S. *The Fourth Medium: How to Use Promotional Literature to Increase Sales and Profits.* Homewood, Ill.: Dow Jones–Irwin, 1986.

Gosden, Freeman F., Jr. *Direct Marketing Success: What Works and Why.* New York: Wiley, 1985.

Hecker, Sidney, and David W. Stewart. *Nonverbal Communication in Advertising.* Lexington, Mass.: Lexington Books, 1987.

Hisrich, Robert D. *Marketing a New Product.* Menlo Park, Cal.: Benjamin/Cummings, 1979.

Hodgson, Richard S. *The Greatest Direct Mail Sales Letters of All Time.* Chicago: Dartnell, 1986.

Joffe, Gerardo. *How to Build a Great Fortune in Mail Order* (7 volumes). San Francisco: Advance Books, 1980.

Jones, John Philip. *What's in a Name?: Advertising and the Concept of Brands.* Lexington, Mass.: Lexington Books, 1986.

Kamaroff, Bernard. *Small-Time Operator.* Laytonville, Cal.: Bell Springs, 1981.

King, Norman. *Big Sales from Small Spaces: Tips and Techniques for Effective Small-Space Advertising.* New York: Facts on File, 1986.

Kuswa, Webster. *Big Paybacks from Small-Budget Advertising*. Chicago: Dartnell, 1982.

Laing, John. *Do-It-Yourself Graphic Design*. New York: Facts on File, 1984.

Lant, Jeffrey L., Dr. *The Unabashed Self-Promoter's Guide*. Cambridge, Mass.: Jeffrey Lant Association, 1983.

Lavin, Henry. *How to Get — and Keep — Good Industrial Customers Through Effective Direct Mail*. Pompano Beach, Florida: Exposition Press of Florida, 1980.

Lazer, William. *Handbook of Demographics for Marketing and Advertising*. Lexington, Mass.: Lexington Books, 1987.

Lesly, Philip. *Lesly's Public Relations Handbook*, Third Edition. Englewood Cliffs, N.J.: Prentice-Hall, 1983.

Lewis, Herschell G. *Direct Mail Copy That Sells!* Englewood Cliffs, N.J.: Prentice-Hall, 1984.

Lodish, Leonard M. *The Advertising and Promotion Challenge*. New York: Oxford, 1986.

Louis, H. Gordon. *How to Handle Your Own Public Relations*. Chicago: Nelson Hall, 1976.

Lyons, John. *Guts: Advertising from the Inside Out*. New York: AMACOM, 1987.

Maas, Jane. *Better Brochures, Catalogs and Mailing Pieces*. New York: St. Martin's, 1984.

Mackay, Harvey. *Swim with the Sharks Without Being Eaten Alive*. New York: William Morrow, 1988.

Malickson, David L., and John W. Nason. *Advertising — How to Write the Kind That Works*. New York: Scribner's, 1977.

McCafferty, Thomas. *In-House Telemarketing: A Master Plan for Starting and Managing a Profitable Telemarketing Program*. Chicago: Probus, 1986.

Miles, John. *Design for Desktop Publishing*. San Francisco: Chronicle Books, 1987.

Nierenberg, Gerard I. *The Art of Creative Thinking*. New York: Cornerstone Library, 1982.

Ogilvy, David. *Confessions of an Advertising Man*. New York: Atheneum, 1980.

Ogilvy, David. *Ogilvy on Advertising*. New York: Crown, 1983.

Ogilvy, David, edited by Joel Raphaelson. *The Unpublished David Ogilvy: His Secrets of Management, Creativity, and Success — from Private Papers to Public Fulminations*. New York: Crown, 1987.

Ortland, Gerald T. *Telemarketing: High Profit Telephone Selling Techniques*. New York: Wiley, 1982.

O'Shaughnessy, John. *Why People Buy*. New York: Oxford, 1987.

Phillips, Michael. *Honest Business*. New York: Random House, 1981.

Phillips, Michael, and Salli Rasberry. *Marketing Without Advertising*. Berkeley: Nolo Press, 1986.

Pope, Jeffrey. *Business to Business Telemarketing*. New York: AMACOM, 1983.

Rapp, Stan, and Thomas L. Collins. *Maximarketing: The New Direction in Promotion, Advertising and Marketing Strategy*. New York: McGraw-Hill, 1987.

Ries, Al, and Jack Trout. *Marketing Warfare*. New York: McGraw-Hill, 1983.

Ries, Al, and Jack Trout. *Positioning: The Battle for Your Mind*. New York: McGraw-Hill, 1980.

Schollhammer, H., and Arthur Kuriloff. *Entrepreneurship and Small Business Management*. New York: Wiley, 1979.

Settle, Robert B., and Pamela L. Alreck. *Why They Buy: American Consumers Inside and Out*. New York: Wiley, 1986.

Siegel, Gonnie McClung. *How to Advertise and Promote Your Small Business*. New York: Wiley, 1978.

Slutsky, Jeff. *Streetfighting: Low Cost Advertising Promotion Strategies for Your Small Business*. Englewood Cliffs, N.J.: Prentice-Hall, 1984.

Smith, Cynthia S. *How to Get Big Results from a Small Advertising Budget*. New York: Hawthorn, 1973.

Soderberg, Norman R. *Public Relationships for the Entrepreneur and the Growing Business*. Chicago: Probus, 1986.

Stansfield, Richard H. *Advertising Manager's Handbook*, Third Edition. Chicago: Dartnell, 1982.

Swann, Alan. *How to Understand and Use Design and Layout*. Cincinnati: Writer's Digest Books, 1987.

Throckmorton, Joan. *Winning Direct Response Advertising: How to Recognize It, Evaluate It, Inspire It, Create It*. Englewood Cliffs, N.J.: Prentice-Hall, 1986.

Todd, Alden. *Finding Facts Fast: How to Find Out What You Want to Know Immediately*. Berkeley: Ten Speed Press, 1979.

White, Matthew. *Turn Your Good Idea into a Profitable Home Video*. New York: St. Martin's, 1987.

Advertising Age, 740 N. Rush St., Chicago, IL 60611.   **Periodicals**

Adweek, 5757 Wilshire Blvd., Los Angeles, CA 90036 (1-800-3-ADWEEK).

Direct Marketing, Hoke Publications, 224 7th St., Garden City, NY 11535.

Entrepreneur, Chase Revel, 631 Wilshire Blvd., Santa Monica, CA 90401.

Harvard Business Review, Soldiers Field Rd., Boston, MA 02163.

In Business, The JG Press, Box 323, 18 S. Seventh St., Emmaus, PA 18049.

Journal of Marketing, American Marketing Association, 222 S. Riverside Plaza, Chicago, IL 60606.

Standard Rate & Data Service (11 directories), Macmillan, Inc., Skokie, IL 60077.

Venture: The Magazine for Entrepreneurs, 35 W. 45th St., New York, NY 10036.

Cluff, Douglas D., and Lesa M. N. Bell. *Power Marketing: 5 Easy*   **Audio**
Steps to Turn Small Advertising Budgets into Big Profits. New York: Adweek Books (1-800-3-ADWEEK).

Dunn, Dick. *How to Write Great Copy*. New York: Adweek Books.

Lazarus, George. *The Marketing Edge: Six Pros Tell You How to Get It* (six interviews). New York: Adweek Books.

Levinson, Jay Conrad. *Secrets of Guerrilla Marketing*. Mill Valley, CA: Guerrilla Marketing International (1-800-621-0851, Ext. 140).

Index